# CLOCKWORK
# FUTURES

ALSO BY BRANDY SCHILLACE:

*Death's Summer Coat*

# CLOCKWORK FUTURES

## THE SCIENCE OF STEAMPUNK AND THE
## REINVENTION OF THE MODERN WORLD

# Brandy Schillace

PEGASUS BOOKS
NEW YORK  LONDON

CLOCKWORK FUTURES

Pegasus Books Ltd.
148 W 37th Street, 13th Floor
New York, NY 10018

First Pegasus Books cloth edition September 2017

Interior design by Maria Fernandez

Library of Congress Cataloging-in-Publication Data is available.

ISBN: 978-1-68177-518-0

10 9 8 7 6 5 4 3 2 1

Printed in the United States of America
Distributed by W. W. Norton & Company

*For Mark*

# CONTENTS

## PERAMBULATION
## The Mad, Mad World

E very good history begins with a *story*. The best of them build bridges through time where one story's end is lost in the next one's beginning, like a dragon swallowing its own tail. This one has its origin in a small book on a high shelf that, through a series of accidents, managed to turn up as the key to a curious set of questions about humans and clockwork, power, steam, and machines. A slim little volume, *The New Epoch* appeared in 1903, but contains a series of lectures given in the last years of the nineteenth century. Its size and antiquity make it practically invisible today; out of print, far from the well-lit shelves of bookstores or even from the bright screens of online purveyors. I only discovered it through someone else's discovery—the author's nephew, Elting Morison, found the book in his uncle's library and worked up his own little volume in 1966, *Men, Machines, and*

*Modern Times*, and exactly fifty years later, a colleague recommended it to *me*.* The chances were slim, the timing contrived, and the general obscurity to which the authors had fallen make the entire episode smack of fantasy and fate. But of course that's how the best stories often begin.

The earth is old; there continues to be some disagreement about *how* old, but far older than mankind, and by extension much, much older than that which mankind *makes*. Technology arrived on this planet, flung from sparks and driven by heat and curiosity and star dust, with all the old means of time and evolution summarily tossed aside. "Within a period so recent that we are practically in the midst of it," says *The New Epoch*, "man has acquired a new capacity, which marks as distinct an epoch in civilization as the earlier achievements made in the savage and barbarous life of primitive society."[1] Once, we sought to control power, to harness the horses, to yoke the oxen, to put the human body to work, In this latest age, humans have learned to *manufacture* power. We've become so used to flipping a light switch that we rarely take a moment to think—really think—about what this meant in our history. But to late Victorians, and particularly late Victorian engineers like George Shattuck Morison, the unassuming author of *The New Epoch*, this was the stuff of dreams. Consider, he asks us: whatever the power of a single machine, that machine can be used to make a better one. Why, the power generated in a Victorian steamship in a single voyage across the Atlantic was enough, Morison estimated, to raise the great Egyptian pyramids.[2] We'd come a long way from the Nile in 1896 (when Morison's first lecture was given), and I turned the pages greedily; I'd discovered a window into a past world and was eager to hear about his designs for the future we now inhabit. Bright with anticipation, certain of ultimate success, and favoring the engineer and maker over the previous century's philosophers and thinkers, the little book seemed to offer the germ of what we today call *steampunk*, that hopeful aesthetic of Victorian future-hunting.

---

\*     Many thanks to James M. Edmonson, chief curator, Dittrick Museum of Medical History.

For aficionados, *steampunk* needs no introduction—but even then, it might need a definition. Nearly every commentary on the subject begins by saying it cannot be defined, that its shape is amorphous and its origin cloudy. Strangely balanced between a nineteenth century that never was and a future that never will be, it's the stuff of dreams, of nostalgia, of alternate pasts and futures that entice with the suave of James Bond and the savvy of Sherlock Holmes. If you think of it as a fiction genre, it can be traced to the work of Jules Verne and H. G. Wells, or linked to the science fiction and fantasy cross-novels of the 1980s and early '90s—like *Infernal Devices* by K. W. Jeter or *The Difference Engine* by William Gibson and Bruce Sterling. Or to the Brian W. Aldiss's 1973 novel *Frankenstein Unbound*, or to Mary Shelley's actual 1818 *Frankenstein*, or to Alan Moore's graphic novel *The League of Extraordinary Gentlemen*, or Philip Pullman's *His Dark Materials*. By being ill-defined, steampunk isn't much troubled by boundaries and limits. It turns up in film and television, at least as early as *The Wild Wild West*, which stirred up an eager following in the 1960s, at about the same time Morison's nephew published *Men, Machines, and Modern Times*. And today, steampunk has blockbuster potential with movies like *The Golden Compass*, Robert Downey Jr.'s *Sherlock Holmes*, *The Prestige*, or even *Van Helsing*, and it likewise turns up in "maker" cultures, among do-it-yourselfers, crafters, and costume designers. We arrive, not at a definition, so much as a composite creature.

Cynthia Miller and Julie Taddeo (editors of *Steaming into a Victorian Future*) describe "steampunk" as a tension between past and future, anachronistic in its technology and magical in its whirring gears and golden machines. Cherie Priest, the acknowledged "queen of steampunk," calls it "an aesthetic movement" based on available tech of the time[3]—before an oil industry, before advanced plastics, before microchips. And with some exceptions, steampunk tends toward the positive; in fact, it's almost a celebration of what technology *might* have been, infused with color and life in a way that coal-bound London never could have been. Pop culture philosopher Professor Henry Jenkins claims that science fiction works by asking questions—probing, prodding "annoying" questions—about the nature of technology. But, he's keen to remind us, steampunk is *not* "Victorian Science Fiction," partly because it doesn't ask questions about the future at all.[4] The point is to look backward, back into that safe past which feels for us like solid ground. Of course, preference for the past has been going around almost as long as there have been pasts to prefer. What qualifies steampunk as a new social experiment, worthy of study? And what makes it important to this unfolding tale of men, machines, science, and power? In the end, it all comes down to the clockwork, and to our often futile attempts to order a mad, mad world.

Imagine the books of China Miéville come to life, a strange combination of *Perdido Street Station* and *Railsea.* Now imagine this redesigned in the vein of *Mad Max.* The extraordinary result might look something like the Neverwas Haul, a steampunk contraption that skirts the Nevada deserts in advance of Burning Man. Roger Whitson, steampunk and Victorian pop culture scholar, describes the Haul as part diesel, part steamship. Its designer and pilot Shannon O'Hare, aka Major Catastrophe, heads up the "Traveling Academy of UnNatural Science." Driven by "Track Banshees," women whose "artisanal skills create vehicles of unsurpassed beauty and power," the wagons literally bring fantasy to life.[5] In an often-cited article by James Schafer and Kate Franklin, "Why Steampunk (Still) Matters," this creative drive turns up as a kind of heroism in an age of faceless technology, an "inherent rejection of disposable consumerist culture and the dominance of our contemporary society by modern day robber barons."[6] To take technology as it is, they argue, steals the gleaming, golden dream and replaces it with something coldly utilitarian. Technology should be about the discovery of a great future, a bold vista, a somewhere-out-there, but it should do so without losing the comforting and familiar vibration of gears we see, rails we laid down, bits and pieces we can get our hands on (and our minds around). Whitson goes on to provide the literary and literal history, in brief, the way "steampunk" was coined by K. W. Jeter in 1987 to describe Victorian fantasies that didn't fit into existing boxes—that amorphous aesthetic we are still trying to pin down. Plenty of lists describe the "call it like you see it" quality of the genre, but there is one thing that seems to be present in all of them, and also in the tinkers, gear-breakers, and costume-play-makers. Beneath, behind, and infusing the mechanism we have the ever present specter of *time*. Time travel, to be sure, but also a preoccupation with time, with futures and with pasts, and with the original "machine" in its own shapeless, measureless housing: the natural universe in its millennia, expanding, contracting, and ever moving.

History tends to work only one way, and most of the time, we don't know when we are "making history." But when George Morison stood at the edge of his century, he was deeply impressed by the movement of the gears beneath him, what they would require, and even something of what they would cost us. The "inanimate manufactured power," as he called it in *The New Epoch*, "is absolutely without sense."[7] It is not moral. This power would "destroy as well as build," and the new civilization would "wipe out the conditions which preceded it."[8] But this destruction was not, in Morison's estimation, "right" or "wrong." It was simply a risk worth taking. I read his words with a growing sense of unease—the little book had started to nag with something like dread. Losses, he concedes, must come. Tribes and cultures and nations may be swept away, in favor of a single over-culture of technology. We must, Morison explains, sacrifice the hardy independence of more "savage" ages for something else—for better food, better clothes, happier individuals, and a time of great peace. Certainly, men may take ill advantage of these new powers, but "the future good of our race lies in utilizing them to the utmost possible extent, and not in trying to retain the good features of conditions which are passing away."[9] The stuff of dreams, all right, but of nightmares too. In fact, this story of technology, clockwork, and power exists precisely at that intersection. On one hand is life and light and possibility—and on the other, death, darkness, dread.

The bubble of Morison's golden dreams casts a long shadow over our time. In his sanguine and optimistic predictions, he reminds us of just how much we can't know about the future. The same shadow of doubt is cast by the words of another man, international, accomplished, and far better known: the eminent French chemist Marcellin Berthelot, in 1894. In April of that year, as guest of honor at the annual banquet of the Chambre Syndicale des Produits Chimiques (Chemical Products Association), he offered a vision of the distant "year 2000." The digression is worthwhile. "The day will come," Berthelot predicted, "when everyone will carry for nourishment his little nitrogen pill, his little portion of fats, his little lump of starch or sugar, his small phial of aromatic spices adjusted to his personal taste—all of this inexpensively produced in inexhaustible quantities by our factories." Chemistry, having solved once and for all the problem of food supply, will have created a utopian world where "mankind will gain in kindness and morality, because he will cease living on the slaughter and destruction of living creatures." The earth will have become "one vast garden, moistened by the effusion of subterranean springs, where the human race will live in the abundance and happiness of the legendary Golden Age."[10] The irony of these words, just decades before World War I, canker our consciousness of technology's "triumph"—cheap goods, better food, and the invention of destruction on a scale our ancestors could scarcely imagine outside of an act of God. Not to worry, say these men of vision: "It is premature to say where the compensation for the loss [. . .] will be found, but we may be sure that such compensation will come."[11] Put more simply: these achievements will be better than we can imagine, and they are happening anyway, so why waste time guessing?[12] But these technologies ushered in death, not life; war, not peace. A failure of imagination at best, and a devastating lack of ethical consideration at its worst. Today, we stand at the launch pad of technological change that Morison and Berthelot couldn't have imagined, and we look over the same radical abyss. We are hunting the future, but the past offers a warning and maybe, just maybe, a chance to redirect. What we need, as the Red Queen in *Through the Looking-Glass* suggests, is a memory that works both ways.* We want to look backward and forward at once, and this is precisely what steampunk fiction attempts to do. (It's no surprise, perhaps, that *Alice in Wonderland* has been so thoroughly "steampunked.")

At its most positive, then, steampunk rebels against the hegemony we've come to associate with technology (the ubiquity of identical cell phones, for instance). It also allows room for stretching imaginative muscle, for looking back and considering whether we may have done otherwise and better. We live in an age of "smart" phones and 3-D printers, where organ tissue may be grown in jars from stem cells, and where some have thought it prudent to attempt a resurrection of mammoth DNA.** We've engineered our cities and reverse-engineered our food (presently, there are plans to begin "growing" plant-based eggs).[13] At the same time, we've arrived at an economic, political, and environmental crisis over our use of oil to fuel this future. The number of articles, books, and programs addressing the horrors of this dependence may only be outstripped by the proliferation of fiction featuring an alternative history: one wherein the combustion engine never reigned at all. Steampunk exists at the intersection of past and present, a playground where steam still powers civilization, but without dampening technological progress. Fabulous fashion, gentlemanly assumptions, and a Victorian style (without the reality of cholera, soot, inequality, and malnutrition that plagued the actual nineteenth century) offer a kind of utopian

---

\*  Elting Morison makes the same connection:

"Living backwards!" Alice repeated in great astonishment. "I never heard of such a thing!"

'—but there's one great advantage in it, that one's memory works both ways."

"I'm sure mine only works one way," Alice remarked. "I can't remember things before they happen."

"It's a poor sort of memory that only works backwards," the Queen remarked.

—"Wool and Water" *Through the Looking-Glass, and What Alice Found There,* by Lewis Carroll.

\*\*  With obvious ethical consequences. Ghosh, Pallab. "Mammoth Genome Sequence Completed," Science and Environment, BBC News, April 23, 2015.

hope—couldn't we have done just as well, the genre asks, if we'd never succumbed to a petroleum economy at all? Or perhaps we might reconsider and reimagine the steps that brought us to the Internet, or to mass transit, or remote control weaponry? The Victorian world charms the postmodern reader and seems far less damaged and damaging than the future upon which these destructive technologies have been unleashed. But a "world war" would never have been possible if not for the very devices, networks, and empires built in the nineteenth century. The American engineer and the French chemist weren't the only ones looking to the future from the cusp of the Victorian Age—others from within it were asking questions, too, and of a piercing and particular kind. Retro-future, a way of looking back to look forward, might just shed light on the juggernaut of the now.

Jules Verne, the French novelist who brought us Captain Nemo's *Nautilus, Around the World in Eighty Days*, and voyages to the moon and the center of the earth, had a profound influence on science fiction. He just as frequently serves as a kind of steampunk forefather, and Verne happens to be Berthelot's exact contemporary. Historian Alan J. Rocke remarks on the parallels: in addition to being countrymen and roughly the same age, both achieved their earliest major career success in the same year, 1863, when Verne published the first of his hugely popular "Fantastic Voyages" and Berthelot was appointed to his major professorship, a chair specially created for him at the Collège de France. But at a time when the chemist toasted the end of poverty and the regeneration of the earth through science, Verne wrote his only truly dystopian novel—*Paris in the 20th Century (Paris au XXe siècle)*. In it, Verne imagines "engines of war" and other machines for replacing humans. Meanwhile, humans themselves have become machine-like; the cruel tycoon Stanislas Boutardin moves "like a piston" through a future that crackles with electricity and clanks with gears. Verne's heroes do not rise like zeppelins through the wreck of this mechanized and unfeeling world. They vanish, crushed under the weight of machinery and left to wander a snowy wasteland, lost poets in a world where the humanities (and the human) no longer have meaning. Verne penned it in the early 1860s, but the publishers refused it; forgotten, it only appeared in print in 1994, but his dark reckoning should serve as a stark reminder that even the nineteenth century was not all tea tables and top hats.

Grim and gritty, without the benefit of child labor laws to protect the innocent or environmental controls to keep us from polluting air and water, the Victorian city housed dark factories and sooty windows. From a medical standpoint, it was a terrifying time of disease and operating theaters that stunk from the offal of dissection, amputation, gangrene. The world, once the province of farmer and field, bounced and clanged on the rails of industrial change. Steam engines, boilers, automatic looms, pump stations, and automobiles brought about radical changes in record time, and human bodies really were crushed under the wheels. If the scientists missed it, the fiction writers did not. In *Hard Times*, English novelist and playwright Charles Dickens describes some of technology's abuses: "It was a town of machinery and tall chimneys [. . .] It had a black canal in it and a river that ran purple with ill smelling dye and vast piles of building full of windows where there was a rattling and a trembling all day long and where the piston of the steam engine worked monotonously up and down like the head of an elephant in a state of melancholy madness."[14] Machinery smacks of chaos and doom. And this fear of technology's advance reflects backward, causing us to reevaluate those earlier devices that once seemed benign. H. Bruce Franklin, editor of *Future Perfect: American Science Fiction of the Nineteenth Century*, describes how this occurs. The Industrial Revolution sees the introduction of bizarre and dangerous creations, and increasing numbers of *fictional* automata too. By 1874, mechanical men were commonplace in novels, as were "vehicles of terror, like the robot in H. D. Jenkins's 'Automaton of Dobello'" of 1872.[15] And in real factories, as in fictional ones, the danger of worker replacement and human injury were very real; the "Luddites" break automatic looms, and fatalities and injuries from boiler explosions and train catastrophes rise. Charles Dickens himself nearly perishes when a train catapults into a ravine; in the days and months that follow, he loses his voice—or, as he put it, "[I] brought someone else's [voice] out of that terrible scene."[16] The Victorian age saw the proliferation of science and technology, but by its terminus had to leave behind the belief that all would go on in good order.

Historian Philip Guedalla described the previous generation as "men in black coats who produced astonishing results while thinking hard all the time about something else"; men who invented railroads while thinking of communications and defense strategies, and who weren't aware that they were by proxy building a brand new world of interconnections (that would ultimately make it as easy to invade as to defend).[17] For all George Shattuck Morison's aspirations for mankind in the latter nineteenth century, the decades he had himself lived through should have been fair warning. They weren't. But by the second decade of the twentieth century, unexamined optimism was far less possible and certainly not advisable. Elting Morison, the author of the *other* little book—and my first introduction to his uncle's *New Epoch*—put it best as a series of questions: *why change? who changes? what design problems? what capital requirements? what industrial modifications?* And most importantly, *who gets hurt,* and *who profits?*[18] The questions aren't new. But the way history answers them is bound up with the very machines we invented, and even today, with the most advanced of our difference engines, the computer, you can only get out what you put in.

Steampunk's fictional worlds try to have it both ways, offering technology via two competing visions. It can be the height of craftsmanship and reason and well-purposed parts; it can heighten and extend the human, but with characteristic Victorian charm. It runs like clockwork. But that same technology also breaks and destroys, tears and rends. Bodies are broken. Jobs are stolen. Death runs rampant. We can point fingers, but that mistakes the first principle: we use (and abuse) technology because someone, somewhere, sometime, *sold us on the idea.* Miller and Taddeo talk about the "magic" of the Victorian era, a time when scientists and medicine men appeared like magicians to control the elements. Heroes and captains and adventurers inspire us in a time of individual invention we can hardly find modern corollaries for—but they *also* worked as carnival barkers who shouted at the entrance of draped tents and offered to sell science for a nickel. Steampunk has been called "fantasy made real," but much of what I've been describing comes from the actual history of actual inventions that actually worked (or didn't, or blew apart in the furnace, or took off a limb or two in the process). George Morison claimed, "The records of the future must be made by men of different types and different habits, [. . .] who will exchange the pleasures and quiet of the university for the roar of the rolling-mill, the buzz of the machine-shop, the obscurity of the mine, the bustle of the railroad."[19] Elting, with the benefit of the Red Queen's memory, revisited that idea fifty years on and recognized invention itself as "a hostile act—a dislocation of existing schemes."[20] Do we, finite and limited creatures in the vault of time under a vast continuum of expanding galaxies, dare disturb the universe? Looking backward to look forward in *history* offers us a chance to do something steampunk fictions, on their own, cannot do: first, we can look beyond the cogs and wheels and see not only how they work but what they are for, what demons they sought to thwart or contain, and what inherent disruption they caused in coming to be. Second, we can see where in this vast arrangement of human activity steampunk has its first origin—not, as some suggest, in the inventions themselves, but in the brilliant and glowing display of them by those who sought to overcome our natural resistance to change. Lastly, we can examine the very fine line between our desire and our dread—for dread and fear are not the same. We can almost always name the thing we fear; it has a face. That which we dread is more uncertain, a sense that something, somewhere, is about to go horribly wrong—a feeling that we cannot prepare for its assault. As far back as Newton in the seventeenth century, the hot-air balloon showmen and medical charlatans of the eighteenth, or the industrial magnates and electrical wonder-workers of the nineteenth century, innovators needed to "sell" science to those who would fund their explorations. . . . And to trade upon the fiction of control. Technology to serve men, not the other way around, they promised—and they often downplayed, ignored, or just failed to see tragedy and consequence along the way.

In other words, we didn't invent steampunk in the twentieth century as a response to today's technology; the "steampunk" ethos has been with science from the beginning, the bright crest of war-works in a battle against our greatest foes: chaos, darkness, privation, anarchy, and death. To *chaos*, seventeenth-century mathematicians sought to bring order; to *darkness*, eighteenth-century explorers and experimenters tried to bring the light. The Victorians fought to bring industry and control to an expanding empire that threatened *privation* and *anarchy*—and at last, the unconquerable enemy that we still fear and face: *death*, the destruction of all our future plans in the fragile, finite human body. Science, and the power of the future, works always to this end: conquer the unconquerable. Showmen and cranks take their place right next to "real" science in this history, bringing the carnival to a waiting public in terror and in wonder. This book reveals a tangled history as much about our dread as about our love of discovery, as much about the train crash as about the gleaming rails. A social history of technology and the seduction of "clockwork futures": this is the story of hope, trepidation, and the struggle of modern science in a steam-powered age. Salesmanship and indeed *showmanship* are the very spirit of steampunk, and the fictions we tell ourselves are part of the scientific framework we have inherited, part of how we "do" science and how we understand its destiny—yesterday, today, and tomorrow.

# PART ONE

"I am much occupied with the investigation of the physical causes [of motions in the Solar System]. My aim in this is to show that the celestial machine is to be likened not to a divine organism but rather to a clockwork . . . insofar as nearly all the manifold movements are carried out by means of a single, quite simple magnetic force. This physical conception is to be presented through calculation and geometry."[1]

—Johannes Kepler

## ONE

## The God of Mathematics

I n the small hours of morning, when sleep has fled and the strange noises of night crowd young imaginations, numbers can be magical. I do not mean math proper, not yet; I mean the solidity and reality of counting. We counted sheep, we counted our toes and small fingers, our elbows and knees—strange preparation for the infinite, uncountable stars that awaited in the velvet dark. I remember feeling small, but with my back against grass and the warm earth under me, I don't ever remember feeling *lost*. The world as I knew it had concrete foundations. I believed it immovable and unshakable. And for many centuries, most of humankind held a similar view. Aristotle claimed as much in 355 B.C.E.: "In the whole range of time past, so far as our inherited records reach, no change appears to have taken place either in the whole scheme of the outermost heaven or in any of its proper parts."[1] Aristotle's universe had no creator; it preexisted all things: an I HAVE BEEN rather than the great I AM. But Aristotelian ideas would be used by Christian Medieval and Renaissance astronomers to build complex mathematical systems for understanding the cosmos as a purpose-built machine. This universal clockwork—the mathematics, even the *numbers*—they claimed, proved the presence of an intelligent creator.

In this book's *perambulation*, I suggested that the earth was old and technology new. David Wootton, a historian used to taking the long view, explains that tool-making humans have been around for about 2 million years, with *Homo sapiens* (our own particular brand of human) arriving about 200,000 years ago, pottery 25,000 years ago, and agriculture between 12,000 and 7,000 years ago.[2] George Shattuck Morison described these "ages" of men, too, from his lecture of 1896. He calls them the "three periods of savagery, followed by three periods of barbarism," with the taming of fire as the initial precondition—fire led man from the first to the second stage, the weapon from the second to the third, and so on until we arrive at written language and "civilization."[3] Despite our long history, written records have existed for only about 6,500 years, and modern technological and scientific innovations of the sort we're talking about occupy only the last 400.[4] Four hundred years: just less than twice the age of the United States Declaration of Independence, a good 125 years after Columbus stumbled into North America, a bare slip of time. Wootton rightly calls the world we live in today "box fresh";[5] Science as we know it was "invented" between the discovery of a new star in 1572 and the publication of Newton's *Opticks* (on light) in 1704.[6] What Morison described as the "new epoch" is the crescendo of a movement forty generations old, heralding the end of "old buildings, old boundaries, and old monuments, and furthermore of customs and ideas, systems of thought and methods of education."[7] Ironically, in *1664* a man named Henry Power claimed almost exactly the same thing: "Me-thinks, I see how all old Rubbish must be thrown away, and the rotten Buildings be overthrown, and carried away with so powerful an Inundation. These are the days that must lay a new Foundation of a more magnificent Philosophy."[8] An English physician and one of the first elected Fellows of the Royal Society, Power felt that he, too, stood on the heaving deck of a brand new epoch. You can only move forward, never back; the wave of change will come. Both men were right about that, but the story can't begin in the shining future with its gleams of promise. It starts in the stench and decay of those "rotten buildings" at the dawn of the seventeenth century, and mankind's firm conviction that *no new knowledge existed*.[9]

Imagine a city, circa 1600: Animals were slaughtered and the entrails and blood left to seep in straw, home to flies and larva and all manner of bacteria. Light came from guttering animal-fat candles, foul smelling and sooty. Close living and no plumbing meant human and animal excrement mixed in streets. Science writer Edward Dolnick describes London as particularly crusted and bleak, but even the palace of French king Louis XIV only cleared its corridors of feces once a week.[10] Today, we may argue over sexing public restrooms, but until fairly recently there were no public facilities, and in fact no *toilets*, at all. Pests were naturally everywhere—from the rat down to the louse. Even "romantic" writing included reference to the ubiquitous flea (most famously John Donne, who attempts to woo by reference to the mixing blood in the insect's gut). Skin disease, rot, and various infections scabbed over the bodies of urban dwellers, rich or poor. Country folk fared only marginally better, their bodies broken under heavy manual labor. War and religious tumult, from the Thirty Years War in Germany to the English Civil War, meant that national governments existed in flux and instability. And to make matters worse, the Black Plague swept across Europe in deadly waves from the fourteenth century to the devastating outbreaks of the 1650s–60s. Medicine stood mostly helpless under the onslaught, and too often the cure killed as fast as the ailment. It could not have been a great comfort to think God was in his heaven and all was right with the world. It was, however, the only comfort going. The educated man before 1600, says Wootton, would have believed that the earth remained fixed and still, and all else revolved around it, and that God was minutely involved and interested in man.[11] Man's only recourse in this deeply flawed but fixed little world was to that same God, whose ways he little understood, but who nonetheless provided the answer to all questions. Even death. The world was just so because God ordained it, and mathematics could prove it.

## *The Mechanics of the Universe*

Nicholas Kratzer arrived in England in 1516, bringing with him the latest ideas in mathematics and astronomy (something that made him a favorite in the court of infamous Henry VIII). A Renaissance man in the literal sense, he believed the "secrets of the universe could be unlocked with precision engineering"—but that's not all. As Dr. James Fox points out in the BBC TV series *A Very British Renaissance*, Kratzer made good on his word. He crafted small, intricate, unbelievably accurate sundials (one with a dial on nine sides), and in so doing "harnessed" the sun itself. Here was the grand design, rendered sensible, miniature, mechanized. And it proved to his eager patrons that all was right and orderly: In the beginning, there was God, and God created the heavens to revolve around the earth. And since Kratzer engineered the dials to be timed for England's particular place on it, the sun rose and set right over king and court as the center of their world. Nothing in the divine clockwork suggested otherwise, and Kratzer's friend, collaborator, and fellow German Hans Holbein further solidified this orderly understanding with his painting *The Ambassadors* [Fig. 1]. Fox, a Cambridge historian with a particular interest in art, takes the time to reveal the work's somewhat mystical symbols: a table with celestial materials on top (many of which were tools Kratzer used in his study of the heavens) and earthly matters on the bottom (including human means of entertainment, like the lute). The painting, like Kratzer's clocks, contained the heavens and the earth, summing up the celestial bodies for human scale consumption. Aristotle and those who followed, like Claudius Ptolemy in 100 C.E., were no longer just men who thought deeply about the cosmos; they were part of a system that, nearly two millennia later, served as the structure undergirding everything, from the king and court, to the religious centers, to men and women out in the muck and mire of daily existence. But the system wasn't without its problems.

If you looked carefully at the night sky, and you did so every night for a year, you would notice something unusual. Stars and planets don't just spin around the visible hemisphere, sliding along the horizon in a predictable way. Instead they appear in odd quadrants at different times of year—and to the careful observer, especially one who believed all of these heavenly bodies were spinning around the earth, the movements would seem erratic and out of order. Given that the precision and permanence of the cosmos offered just about the only succor in the dying world, the anomalies had to be figured into the system. The astronomer Ptolemy tackled this puzzle in the second century. Taking it as unquestionable truth that the sun went around the earth, he theorized planets must move in something called "epicycles" or miniature circles while on the big loop or "deferent" around it. But an intrepid and stalwart star watcher might notice something else amiss too. Planets and stars appear to *speed up* and *slow down*. Now what? Ptolemy wasn't discouraged. He decided that half the epicycle runs *counter* to the deferent, which makes it appear that things change speed or reverse direction. He even took it a step further by inventing a point called the "equant," meaning Ptolemy moved the observation point explicitly off the center to account for variation. If that sounds confusing, it should. The system worked on mathematical principles that are not terribly easy to explain. You *can* model them, however, and Giovanni de' Dondi created a complex gearwork called the astrarium in Padua in the 1380s to demonstrate exactly how the solar system, as Ptolemy understood it, functioned. In order to answer the need for mathematical precision, Ptolemy gave rise to the concept Kratzer would later take for granted: we live in a universe that operates like a clock. The mathematics was correct. But the model, with its assurance of a constant and predictable universe, was *wrong*.

In 1572, a new star appeared. The discoverer, Tycho Brahe, used trigonometry to show that what he saw through his telescope, a brilliant star that could be seen with the naked eye inhabited the heavens, not the corruptible lower spheres.[12] Who among us would be so lucky? New stars are rare phenomena—an earthbound explorer might only view one in a thousand years. But for it to appear in the "unchanging" cosmos at a time when demons still caused ill winds and witches were still burned (along with those who challenged deeply held views), this *nova* signified terrifying portents. Why was it there? *How* was it there? Was the world ending? Brahe spent the next fifteen years observing the heavens and measuring the immeasurable, though he never entirely questioned that the Earth remained stationary. That was left to another: Italian astronomer, mathematician, and engineer Galileo Galilei. The ill-fated Galileo wasn't the first to see the flaw in the Earth-centered design, just the first to run afoul of the Inquisition (a series of offices of the Catholic Church charged with combating heresy and prosecuting heretics). Renaissance mathematician Nicolaus Copernicus made arguments against Ptolemy in the 1500s. He published *De revolutionibus orbium coelestium* (On the Revolutions of the Celestial Spheres) just before his death in 1543 and ushered in a true "paradigm shift"—a radical shift in thinking that some have credited with the dawn of the "Scientific Revolution." But the new system did not make waves; it didn't even capture the attention of the Church until six decades later, when Galileo was condemned for holding Copernican beliefs. Timing, as they say, is everything.

Suggesting that we weren't at the center of the universe and that years of careful modeling had to be thrown out the window wasn't winning Galileo any friends, but it had less to do with stars and planets and more to do with heresy—or contradicting the Church and its doctrine. Galileo's problems begin in 1611, a full sixty years after the death of "scientific" clockmaker Nicholas Kratzer. They end with his formal interrogation in 1633 (for eighteen days), the threat of torture, and house arrest, under which he died in 1642. The contest is often reduced to religion versus science, but it's actually far more complicated and more interesting. The cosmos, like the nine-sided sundial, supported the unquestioned mechanism of *order*, but also of *authority*. The Inquisition expanded its scope of influence in part to assert its authority over the Protestant Reformation, which began when Martin Luther nailed his "Ninety-Five Theses" to a cathedral door. A few years later, Henry VIII asserted his own authority by abolishing the Catholic Church in England and becoming a rule unto himself. The struggle over who has the power to give orders and make rules informs the backstory of all the conflicts to follow, including the English Civil War, which cost King Charles I his head and ushered in Puritan rule, which itself ended in violence—some of it posthumous (Oliver Cromwell was exhumed so he could be hanged). Regardless of personal faith or conviction, the 1600s would have seemed out of order indeed, nothing stable, nothing secure.

We fear chaos. We should. It speaks of confusion, disorder, mayhem. Chaos seeped from diseased bodies and sloughed from rotten timbers, but with Galileo's ideas came a chaotic threat to the very cosmos above, the dwelling of God, and all the systems that had been built upon their unchanging nature. Today, with our general understanding of a constantly expanding universe, with our quantum physics and string theory, our acceptance of earth's insignificance in the vast reaches of space, it's hard to imagine the earth-shattering effect of new knowledge. But consider again what it meant the first time you heard it, when you wrapped your child-mind around the idea that Earth hurtled through space at a 1,080 miles per hour—or that the sun, which appeared to *rise* and *set*, actually stood still as we spun around and around. Imagine the foundation of belief crumbling from beneath you, and not only you, but a whole generation. When the Copernican system finally supplanted the Ptolemaic, it required new visions, new mathematical models, new means of reproducing that original desire for precision. In the process, "magic numbers" took on new significance, and empirical evidence new meaning. But something else emerged at its very birth, and from the same grisly throes. A dark twin, a *doubt*: a dread that the world had no foundations after all, that even religion may be faulty and untrustworthy, crept forth with the first questions of the empirical mind. Who would order the universe afresh? The coded language of the cosmos would be deciphered by three men, independently: Johannes Kepler, Isaac

Newton, and Gottfried Wilhelm Leibniz. Together, they imposed an order never thought of before, making the heavens and the earth alike mathematically legible. The rest of this chapter follows these unlikely heroes of "new philosophy" on their quest to find order in the numbers, and a divine God within the Machine.

## Of Gods and Machines

*"Grandfather Clock is a creature of logic and precision [. . .] He allows neither change nor error."*
—S. M. Peters, *Whitechapel Gods*

In S. M. Peters's modern steampunk novel *Whitechapel Gods*, the world isn't running *like* clockwork, it's run *by* clockwork—by Grandfather Clock, an austere, joyless, rule-keeper of logic and precision, and Mama Engine, whose creative, reproductive power is also fiery destruction. Oliver (the vaguely Dickensian hero of Peters's novel) struggles against the Whitechapel Gods, but also against his own physical nature. A kind of steampunk loom-breaker, Oliver intends to bring the system down by use of his own body. The scene screams at us, visceral and violent: wires and spikes pierce Oliver's flesh and coil under his ribs. This is communion with the "father," and he "succumbed without struggle, though the relentless pounding wounded him in ways he did not know he could be damaged."[13] Oliver's great achievement is also his great fear—he loses awareness of himself, even as he destroys his god in the wasteland district of Whitechapel.

Peters gets a great deal right about Victorian London. Grit and bleakness, the harbingers and consequence of industrialization, smudge each page. The novel begins with a quote by Arthur G. Morrison from 1889, who described Whitechapel during the time of Jack the Ripper as "a horrible black labyrinth [. . .] reeking from end to end with the vilest exhalations" and "swarming with human vermin, whose trade is robbery and whose recreation is murder."[14] And of course, there is the Ripper himself, death stalking the alleyways. Arthur Morrison compares London to Rome—not the well-ordered columns we tend to imagine, but Rome at its fall, decadent and dangerous. But the "oozing" streets of 1889 London might also be the excrement-crusted alleys more than a century earlier, when Jonathan Swift published "Description of a City Shower":

> Filth of all hues and odors seem to tell
> What street they sailed from, by their sight and smell.
> [. . .]
> Sweepings from butchers' stalls, dung, guts, and blood,
> Drowned puppies, stinking sprats, all drenched in mud,
> Dead cats, and turnip tops, come tumbling down the flood.
> (ll. 53–56, 61–63)

Like many steampunk novels, *Whitechapel Gods* looks backward to the Victorians. We are used to thinking of technology and industry as beginning somewhere in the nineteenth century, but the roots of mechanism, application, engineering, and mathematics begin in the muck of the seventeenth century. Henry Power had it right; something was coming that would dramatically reorder everything, but that something required the city itself, with its consolidation of men and of power and resources. In *Whitechapel Gods*, the bounds of urbanity self-expand; city and mechanism "grow" wild like kudzu, into everything. Even people. Its "maze of beams" sprout like living things, creeping ever upward. But the real seventeenth-century city was likewise a generating engine, an idea machine: London, heaving in filth, gives birth to pristine ideas. And in some ways, pestilence itself helped science on its journey. As a result of the plague years, a young man with a lot on his mind would change the world forever. Isolated on an

11

obscure family farm, Isaac Newton discovered calculus, the language for decoding the mysteries of God and discovering the order of his universe.

Myths have great power, and myths about men are of an equally deep and clinging kind. Today, we consider Newton a great rational figure, a first father of science who did away with irrational ideas by establishing grand order, impregnable laws, and unshakable mathematics. But Newton didn't use math to turn religion on its head or to banish superstition; his aim was to *prove* God's existence and to commune with the divine. Mathematics, science, technology were not considered ungodly. His ideas so influenced the eighteenth and nineteenth centuries and their skepticism that we forget he was, in fact, a near contemporary to Galileo and an ardent believer in God (though he, too, believed the earth went around the sun). Ptolemy may have established the gears of the universe as early as 100 C.E., but the concept of a clockwork universe didn't specifically interest Newton. Quiet, solitary, unusual in his habits, and unique for his pinpoint focus, Newton was a code-breaker, not a clockmaker. To understand him, this man who sought to know the secrets of the heavens and the earth, we must first understand his inspiration. From our vantage point, Newton appears to stand apart, a lonely divinity of scientific reason. But Newton's world wasn't empty; the first scientific revolution had already begun, and, as he famously stated, he could stand on the shoulders of giants . . . including among them Galileo himself.*

Galileo may have ended in bad straits, but his career took up a much broader territory than just the Copernican dilemma. He was a gifted mathematician working on concepts of gravity, and proved that all objects (regardless of weight) fall at the same rate in a vacuum. Publication was the key to success, and Galileo understood, perhaps better than anyone at the time, that making bold claims required speed and presumption.

---

\* Letter to Robert Hooke (February 15, 1676). However, considering their strained relationship and exactly what Newton was claiming, it would be a misinterpretation to consider this humility at its best.

As mathematics became the language not only for understanding the heavens, but also for the earth, it opened up whole new realities—and larger scope of influence. "No human investigation can be termed true science," Leonardo da Vinci claimed in his *Treatise on Painting*, "if it is not capable of mathematical demonstration." Math had once been thought a rather "mechanical" occupation (associated with grease and labor); now it approached the heavens. In 1609, Johannes Kepler made math his grand obsession; not only did he work out God's mystic numbers in a complex blueprint, he also wrote *fiction* about it—a voyage to the moon long before Jules Verne conceived of any such thing.

In *Men, Machines, and Modern Times*, Elting Morison describes a "syndrome" that afflicts innovators, inventors, and solitary men of genius (and in the period before women were allowed an education or many other rights generally, the records do favor *men*). A surprising number of those who brought about great changes drank a good deal, were careless with money, and had troubled relationships.[15] They were loners. They got kicked out of school. They mostly annoyed and badgered those around them, the subject of as much vicious and bitter rancor as of adulation. They fought systems, they rebelled against the status quo, and they generally had trouble making themselves understood. But the reason we've heard of these "heroes" involved one further characteristic. They were *loud*. Convinced of their own superiority or of the superiority of their ideas, they rushed to publish, even when publishing might be dangerous. Galileo's pamphlet *Starry Messenger* about Tycho Brahe's new nova acknowledged no one else but Copernicus and himself (something even Kepler thought was a bit unfair).[16] Galileo hurried to publish because he knew others might be onto him. The star had been seen by much of the known world, and in fact a man named Thomas Harriot was hot on his heels. Discovery by "reason," Galileo argued, "is like racing and not like hauling, and a single Arabian steed can outrun a hundred plowhorses."[17] Few people have heard of Harriot because it pays to be first—to be flamboyant, insistent, incessant—and, in Johannes Kepler's case, obsessed, and possibly a bit unhinged.

Like Newton and Galileo, Kepler spent most of his life working over the heavens, but his start was far from stellar: premature, stricken with smallpox, chronic sores on his feet, and separated from his closest associates by a genius intellect and a bad temper.[18] (According to Elting, the perfect concoction for greatness.) Kepler found his stride at the early age of twenty-four. It might be described as a religious experience; Kepler felt he'd discovered the secret truth behind the Copernican, sun-centered system, and in so doing, he had communed with his God. In mathematics, we are often asked to provide *proofs*. From Aristotle onward, the object, however, was to arrive at *a priori*. We have our own meanings for this, today, but for Aristotle, it meant: "reasoning from the one, true cause to an effect."[19] In other words, we begin with an assumption (assuming the sun goes around the earth, for instance) and then using this assumption, we ferret out the details. It's easy to see why the mathematics of Ptolemy became so incredibly complicated; he was attempting to fit the math to the system he already assumed (wrongly) to be true. Tycho Brahe had his own stumbling block of assumption; he may have abandoned the original Aristotelian idea, but he never let go of the "truth" on a stationary earth. It's a caution that even today many hold beliefs that they can accidentally bend theorums to fit. Kepler believed the earth moved around the sun, and he desperately wanted to know *why*. God did not leave anything to chance, and so, Kepler reasoned, there must be *rules*. The opening of Kepler's first publication, *The Sacred Mystery of the Universe*, wastes no time is asserting its grand design: *Quid mundus, quae causa Deo, ratioque creandi*. Kepler himself would discover "what the world is like, that is, God's cause and plan for creating it."[20]

Kepler began his quest like an alchemist, manipulating numerology. Jupiter and Saturn orbited with a conjunction (a point where they aligned) at points that were 117 degrees apart. In other words, they would line up every time 117 degrees from where, on the zodiac circle, they appeared before. We can imagine it as a pie, a circle with a point at the top and then one on either side, as though an invisible triangle sat among them. Now imagine Kepler, alone, leaning fervently over a drafting table and marking one conjunction arrangement after another until he arrived at a circle with multiple dots, evenly spaced. Kepler's awe blossomed from his slate and chalk, his heart thundering as connecting the dots formed beautiful triangles, and at the center of them, a new circle. A design, merely, but it offered Kepler a glimpse of heaven's own language. There were six known planets; each of them had a shape in space—and the relation of their orbits in three dimensions corresponded, Kepler reasoned, to three-dimensional polyhedra (solids with flat polygonal faces, straight edges and sharp corners).[21] Certain that he'd pulled back a curtain and revealed the cogs and gears behind our reality, Kepler struggled to put his discoveries into a language others would understand. His diagrams grew, circles and triangles nestled in squares, then pentagons, then hexagons, each with a circle between. A bit like Kratzer's sundials, he'd developed a system where each planetary orbit had its own shape, and all shapes could be mathematically presented. Kepler's architecture of the universe was elegant, simple—but the numbers refused to add up. Desperate to save so elegant a theory, he suffered endless computations, and concluded with a model that fit every shape into a three-dimensional model, a soccer-ball universe, of sorts. To Kepler, it was beautiful: "No one ever produced a first work more deserving of admiration, more auspicious and, as far as its subject is concerned, more worthy."[22] Kepler's book catapulted his career in astronomy, and he apprenticed to Tycho Brahe on the strength of his endeavors. Brahe offered the chance at delving yet deeper into God's mysteries: offered, but didn't entirely deliver. If Galileo proved the necessity of getting work first into print, Brahe was a master of the bluster it took to keep on top—and he guarded his secrets carefully. Kepler only attained them at his death, and then spent twenty years trying to crack, not *God's* code, but Tycho's. What he found opened a door in the world, a vault out of which "truth" came with yet more chaotic doubt. The geometric model and all its gleaming, angelic order was wrong again. But worse, the holy shape of the circle, a sign of perfection from time immemorial, was also wrong.

Kepler knew when to throw out systems that didn't fit the data. That alone makes him unique—we have trouble doing that even today in our own lives. But consider that Kepler began his training as a theologian, that he was a mystic who believed in a magical code behind all things, and that his faith remained unshaken and unshakable throughout all his life. Kepler (and many of his contemporaries) believed geometry had been "inscribed on the human soul" when it was created.[23] God meant humans to discover his mystic language, and Kepler ardently believed that mathematical truths, like moral laws, were part of mankind's natural allotment, and also a sovereign responsibility. Kepler felt it was his duty to discover God's plan for the world, and in a world of chaos, his commitment actually approximates a kind of stoic heroism. If there were no geniuses willing to take on this momentous task, Kepler would do it, to the glory of God. He does not fail in his designs, except in one way—in pressing for "true" answers, he accidentally finds them, and they are not at all what he expected. What Kepler ultimately contributes are his three laws of planetary movement, each one arriving like a fever dream on waking and captured through messy, deliberate, painful equation-making: that the orbits of the planets are ellipses and not circles; that the radius vector from sun to planet sweeps out equal areas in equal times; and that the square of a planet's cycle period divided by the cube of its mean distance from the sun is a constant.[24] That last one is the kicker; it offers a glimpse at exactly what kind of mental acrobatics Kepler was capable of sustaining—in brief, you choose one planet, multiply its orbit by itself three times, square the planet's year, then divide the first answer by the second. Never mind that most of us don't have a reason to try: imagine the mental dexterity required to discover and refine it. But this, Kepler asserted, was the language of God. And thus all the more worth speaking.

## *In God's Own Tongue*

Nineteenth-century theologian George MacDonald speaks of heaven as a place where "all that is not music is silence." Kepler spent his later years exploring math and mysticism, his genius consumed with numerology as he sought the distant harmonies of God's vernacular. He died in 1630, his grave marked with an epitaph he penned for himself: "I used to measure the skies, now I measure the shadows of Earth/Although my mind was sky-bound, the shadow of my body lies here." Death comes, but though Kepler's body may never have left the stratosphere, his claims about the cosmic realm of his mental faculties are only slightly exaggerated. How else does man first get to those heavens if not by flights of fancy? One of Kepler's first such "trips" is chronicled in the *Somnium*, or *a dream of lunar astronomy*. Born out of Kepler's early interest in lunar geography (such as mountains on the moon), *Somnium* wouldn't appear in print until three years after his death. Related as a "dream," partly to get around potential claims of heresy, and partly to gloss over the enormity of what he attempts to describe, the book takes Kepler (recharacterized as young Duracotus) to the surface of the moon.*

As in the tradition of most science fiction, Kepler deals with the journey not "magically" but with complex discussions about where to land, how to handle the trajectory and orbits of both heavenly bodies, and even the difficulties of breathing on the journey. Duracotus and his guide hide in a cave on the lunar surface to protect them from the sun, and discover from the inhabitants that Lavania, as the moon is called, has two hemispheres: Subvolva and Privolva. Subvolva because the Earth (Volva) is always *above* the moon, and Privolva which never glimpsed the earth at all.[25] Considering the massive climate shifts between dark and light sides, Kepler gives Lavinia mountains, caves and crevices, and bodies of water for the residents to hide in; he also gives them exaggerated size and short life spans, extrapolating from his understanding of the natural world on earth, but not replicating it. Kepler never lived to see it in print, but when it appeared, originally in Latin,

---

\*   *Somnium* has an elaborate frame narrative where a "wise woman" sells her son Duracotus to Tycho Brahe in a fit of anger, only to recover him sometime later, when he has conveniently been taught all about astronomy under Brahe's tutelage (like Kepler himself). Upon discussion with his mother, however, the hero discovered that it is possible, with the assistant of a Daemon, to go on a Lunar voyage and test the theories he has mastered.

few knew what to do with it. Like Tycho Brahe's *nova*, the hovering star of Kepler's "dream" offered something really and truly new: a mathematical mind turned loose on imagination, attempting to build worlds most could not imagine in a language that melded new philosophy (science), mythology, and faith. We would recognize it today as "science fiction." It was the first time an astronomer would seek out (in fancy as well as fact) the geography of the moon—first, but not *last*.

A century and a half later, William Herschel, a British astronomer of German origin and the inventor of the most advanced telescopes of his day, published "Observations on the Mountains of the Moon" in 1780. Half scientific treatise, half dreamlike musing, Herschel's extravagant claims keep pace with Kepler's little book: "[Perhaps] *the Moon is the planet and the Earth the satellite!* Are we not a larger moon to the Moon, than she is to us? [. . .] What a glorious view of the heavens from the Moon! How beautifully diversified with hills and valleys! [. . .] Do not all the elements seem at war here, when we compare the earth with the Moon?"[26] Herschel does not make the journey there, but he does discover a new planet, and as with Brahe's discovery and Kepler's mathematics, the "new" turned the "old" on its head. Again. Everything men understood in the eighteenth century about stability and creation was about to shift—but that assumes a stability to start with. In other words, by Herschel's day, plotting the moon's geography no longer seemed impossible. Far from it. In the age that followed Kepler, measuring the cosmos and measuring the natural world made equal sense to most educated people. Despite the chaos of the 1600s, despite the plague and disaster, despite the mad upheaval of the heavenly cosmos in the hands of Copernicus and Galileo (and the more mad response of the Church to either), by 1780 educated people took *order* for granted. Order was light in the darkness, life from death and decay. Galileo and Kepler had proven that God spoke in numbers, but for a full triumph of order over chaos we return again to the lone thinker on a farm in the year 1665. Newton would countenance no competitors.

History remembers Newton as a beacon, an unearthly figure capable of feats no other had achieved before—and possibly since, at least until

Einstein. Newton was inclined to agree. Born during Christmas,* he felt that his own birth portended "apartness" from others; God had ordained him, destined him, for greatness. And God, as Edward Dolnick puts it, did indeed seem to "whisper secrets" into Newton's ear.[27] Though, if we look at Newton's methods, his persistence, his dogged determination to accept no seconds, we see something else. Johannes Kepler's life burst forward in fits and starts of inspiration; he grasped God's plan as one captures lightning in a jar. Newton, by contrast, did not wait for divine inspiration; he pursued her and ran her to ground. He would align himself to a problem with laser focus, and he would not let it go—not even for bodily necessities of food and water and exercise. Newton believed the Bible's code had been reserved only for "a remnant"; like Galileo before him, he did not suppose that this knowledge belonged to the masses, but to the chosen. Newton's fondest phrase comes from Isaiah: "I will give thee the treasures of darkness, and hidden riches of secret places." The fact that Newton sought them out, more like the adventurers and explorers of the next century than the quiet thinkers of the previous one, never diminished in his own mind the nature of the "gift." What God and Nature do not give easily, Newton himself would grasp with a tenacity that surpassed his contemporaries almost as often as his intellect overmastered them.

Newton's fervent nature and his dedication to mathematics does not disqualify in the least his belief in a Creator-God. In fact, he turned the Latin version of his own name into an anagram: *Jeova Sanctus Unus,* or God's Holy One.[28] There may be reasons beyond the convenience of Latin (and his belief in his own supernatural intelligence) to choose *Jehovah* instead of *Christus* to name the Divine. Newton did not believe in the Trinity, but rather that God the father and Jesus the son were entirely separate entities; the three-in-one he considered a mathematical absurdity, and he nearly ruined his career as Lucasian Professor of Mathematics at Cambridge's Trinity College as a result.[29] Famed Newton historian Betty Jo Teeter Dobbs explains his ardent

---

\* On January 4, the "11th day of Christmas" in the Gregorian calendar.

belief in unity (rather than Trinity) as "a way to reunite his many brilliant facets, which however well-polished, now remain incomplete fragments."[30] There was room for only one singular God in Newton's math—a recipe of parts joining in mystic unity of truth. If Newton's mystic conscriptions sound like the convictions of alchemists, they should: Newton practiced alchemy from his youth. We are used to thinking of the alchemists as wizards turning lead into gold, and it's true that this did remain a preoccupation for centuries. However, alchemy also serves as the origination of chemistry, and even the earliest recipes from Egypt sound more like lab applications than witch's brew: "Lime, one dram; sulfur, previously ground [. . .] add sharp vinegar or the urine of a youth; heat from underneath until the mixture looks like blood. Filter it from sediments and use it pure."[31] While the foul-smelling solution that results likely served in the search for transforming metals, Newton's applications had more to do with his own body. He mixed potions from turpentine, rose water, beeswax, olive oil, red sandalwood and more, and he recommended it internally for "consumption" and externally for "green wounds."[32] He did not think that man could be made divine, nor could he grasp immortality from his bizarre exploits. But Newton did want to preserve the sacred orderliness of the body for as long as he could. Self-experimentation continued well into adulthood as he began studying the principles of light, most famously when he stuck a pin or "bodkin" between his eye and eye bone. It may seem rash—and considering he may have killed himself off before his greatest discoveries, it very likely was—but Newton had a method. He believed in inspiration, but of a kind that required deep attention and observation. It was the mind, and not the body, that approached God.

The only reason, Newton maintained, that no one had yet discovered the secrets of God (not even Kepler) was that no one had yet been good enough or smart enough or "chosen" enough to do it. Newton addressed himself to Nature and to the bounds of the earth through physics and alchemy; he addressed himself to the Divine and to the Heavens through a mathematics devised and developed for the sole purpose of understanding the actions of Providence. It wasn't as simple as seeing God in the clockwork—and Newton does not expressly refer to a clockwork universe. All the same, his quest was for "the simple machinery through which God creates, governs, sustains and replenishes."[33] The simple machinery, however, turns out to be remarkably complicated in application, and it required God's constant maintenance or risked collapsing inward into nothingness. He was "a God of order and not of confusion," Newton maintained, and yet in his understanding of the corruptible universe, Newton also encouraged an idea of entropy. Chaos threatens *always*. We need God *always* to establish or reestablish order, and we need a man like Newton to find his laws and to explain them. For that, Newton didn't need alchemy; he needed a new kind of math. And so, by "thinking of it continually," he invented calculus.

By Newton's own account, the time between 1665 and 1666 was a year of "magical thinking." He published the order of events late in his life, mentioning almost casually his progress into an idea that changed science and physics forever. It makes for some convoluted reading, but it's worth the effort if only to see the master displaying his treasures:

> In the beginning of the year 1665, I found the Method of approximating series & the Rule for reducing any dignity of any Binomial into such a series. The same year in May I found the method of Tangents . . . , & in November had the direct method of fluxions & the next year in January had the Theory of Colours & in May following I had entrance into ye inverse method of fluxions. And the same year I began to think of gravity extending to ye orb of the Moon & (having found out how to estimate the force with wcl globe revolving within a sphere presses the surface of the sphere) from Kepler's rule . . . I deduced that the forces which keep the Planets in their Orbs must [be] reciprocally as the squares of their distances from the centers about will they revolve: & thereby compared the force requisite to keep the Moon in her Orb with the force of gravity

at the surface of the earth, & found them answer pretty nearly. All this was in the two plague years of 1665 & 1666.[34]

Newton's definitive biographer, Richard Westfall, calls this passage "the foundation of the myth of Newton's annus mirabilis, his marvelous year of discovery."[35] Tycho Brahe may have been a showman, and Kepler a mystic, but Newton is the Hero of Science. He has all the qualities that make for Elting's "men of genius": an auspicious and mysterious beginning, born on a fateful date and to a world that clearly needed his clarity and intellect. A loner, irascible, distant, and yet domineering, Newton arrived as a man set apart, a man with an origin myth, and most importantly, a man with a quest. He set out to invent a new language for translating mystery to math, and then to apply that math to the universe around him. It's hard to miss the importance of calculus, but let's recall that the seventeenth century world could not, before Newton, answer even the most basic questions: Why does an apple fall? Why does a cannonball travel in an arc? How fast is it moving? Can I predict when it will arrive? The ubiquity of our technology allows us access to the greatest discoveries of the greatest minds in seconds; we have invented machines to calculate *for* us, but Newton was, himself, a calculation machine. Everything that follows, invention, mechanics, engineering, physics—the launching of satellites and the landing of spacecraft, and more importantly, the ability of one sphere spinning at 1,040 mph to launch something and hit another (spinning at 539 mph) with a Mars rover the size of a Land Rover—comes of this one, incredible discovery: calculus, the language of God.

Newton had a rare capacity for *ecstasy*, what Richard Westfall describes as "stepping outside of himself and becoming wholly absorbed in the problems on which he worked," forgetting to sleep and eat.[36] But Westfall also points out the fact that Newton's private papers do not agree with Newton's year of magic. His achievements remain mind-boggling, certainly. But that he found the answers easily, and in so short of a time, tells only half the tale—and this, by design. Newton's published narrative of events changed his achievements from human endeavors to superhuman feats. The journals, however, point to years of thinking, of sustained inquiry, of mistakes and false steps, of forward and backward motion as well greased as any piston. This more correct version emerges from history, but it makes for a far less interesting *story*. Newton might not have acknowledged his debts to those coming before him, but he'd surely learned from Galileo (who hadn't acknowledged them either) that the way to greatness involved mythmaking. Newton had prophesied his own godlike greatness, and he made sure to tell the story that best revealed that truth. The trouble was that he started too late. The hero had a rival.

Newton believed he alone had been tapped to reveal that God's universe operated in simple elegance, a mathematical structure that could be extrapolated from the heavenly bodies to earthly ones. Gottfried Leibniz, a German polymath fresh from the excitement and turmoil of European philosophy (and pointed in every direction at once), had rather different plans. Leibniz was a rationalist, not a mystic: "nothing in heaven or on earth, no mystery in religion, no secret in nature" defied the power of human reason.[37] Rather than considering cosmic principles as secret treasures of the few, Leibniz felt God's laws were intended for everyone, or at least as many as would turn their attention to mathematical study. To Leibniz, the world was universally good—mathematics existed to prove it could not have been otherwise. He did not see chaos as an enemy to be defeated. Given the unbelievable breadth of his interests and the hectic way he pursued them, he probably didn't see chaos anywhere at all. If Newton was a solitary stoic, Leibniz was his darker angel; attractive, charming, social, and visionary, he didn't make conversation so much as high-velocity impact. Dressed in silk stockings and curls and newly arrived from the French court, the savant conquered mathematics along the way to more elaborate schemes, nearly falling into the works that Newton labored over. In the midst of plans for "a museum of everything" that would have rivaled the carnivals and pseudoscientific sideshows of the Victorian era with its jumble of telescopes, tightropes, tumblers, and fireeaters—Leibniz discovered Newton's discovery. He'd found calculus, nearly two decades after Newton, but *before* Newton published his work—and he found it, like the "little men in black coats" by "thinking all the time of something else."

Leibniz's thoughts ricocheted a bit like the projectiles Galileo spent so much time measuring, but one theme persists—and would recur in the grim end of his career: the calculation of *conflict*. He wanted to build a machine for turning philosophical disputes into a numeric language. Problems could be fed into the design and the solutions rendered without recourse to argument: logic prevailed because the numbers, he theorized, wouldn't lie. "If controversies were to arise," he suggests, philosophers could kindly invite one another to "calculate." He even theorized a means of translating everything to the very simplest forms—to 0s and 1s—which is precisely how our computers work today. That doesn't mean Leibniz didn't value words themselves; historians lament that though an editorial team is right now trying to turn over 100,000 of his manuscript pages into a collection, they despair of finishing in their lifetimes.[38] He valued discourse. But when Leibniz turned over and over the problems and solutions in his mind, he conversed with the greats—reading René Descartes's complex geometry as though it were a novel rather than the long terrible slog that Newton (and most of the rest of us) found it. But there are riddles, still. For one, how did all of this result in two simultaneous and independent discoveries of calculus? It's a bit like Douglas Adams's quandary in *The Hitchhiker's Guide to the Galaxy*. The "Answer to the Ultimate Question of Life, the Universe, and Everything" turns out to be the number forty-two. The interesting bit isn't the answer, but the question.

The riddle that puzzled the masters had been infinity. Calculus revealed for the first time how speed, distance, and acceleration were linked; it helped you get from one to the other, to measure the previously immeasurable, with an equation. To put it more simply, as Dolnick does, "questions about bests and worsts," comparison about whether a quantity was at its maximum or minimum, could be readily and easily answered.[39] That's a bit like measuring something's speed after it has ceased to move. Much of our world isn't about stopped motion, however; it's about acceleration, about how speed changes over time. The frenetic discovery of quantum physics, of $E=mc^2$, of the speed of light that never changes, and a galaxy where everything else, including space and time, *does*, could never have been possible if not for this: *How do objects fall, and how fast do they travel at each infinitesimal point along the way?* Calculus "was a device for analyzing how things change as time passes."[40] But if we look closer, we see that the Great Simple Truth of calculus strikes a killing blow at the very order Newton, and even Leibniz, thought they were defending.

## *Killing the Gods*

I want to return, for a moment, to the dark void of night we've all faced at one point or another in our lives. Growing up, as we have, in an age where the exposure of nature's secrets is a necessary part of education, we can miss the significance of the great and wondrous *beyond*, the mystical realm of gods and demons and sorcerers. Steampunk aficionados resist being considered "fantasy" or "escapist," but that prefigures these as negative terms. For Newton and his contemporaries, "science" infused natural philosophy with mystery and made room for flights of fancy. Kepler wrote fiction, Newton practiced alchemy and apocalyptic code-breaking, and Leibniz believed that dogs might be trained to speak. They all lived with contradiction, believing that God ordered the world and spoke in mathematical proofs that we could decipher and use to study the natural as well as the celestial world, and *also* believing in a great host of things we today would call fantasy and superstition. Like them, we have both a desire for and dread of the unknown. Because of them, we believe those secrets are discoverable, and while we might not think in terms of magic numbers, it's the numbers that made this possible. Calculus, like Oliver in *Whitechapel Gods*, is the key to the machine. And in the same way, calculus reduced supernatural to natural, and gods to men. When Newton and Leibniz battled over the math, they poked critical holes in the mythology of discovery. Newton would win the war—but Leibniz rent holes in Newton's careful theorems by introducing a cloud of doubt. Newton argued that the world was like a machine, and God was the master mathematician. But if the world worked perfectly, ordered by numbers and stalwart equations, then *it didn't need a God at all.*

The battle began with clear lines; each man claimed to have discovered calculus first, and independently, and neither could believe it possible that *another* genius had come up, separately, with the same thing.* Another clear difference had to do with style. Leibniz aimed to make calculus available to anyone, even to the "unworthy" and "unchosen" souls that Newton disdained. Newton's failure to publish probably stems from just that: it was his duty to discover God's order, not to let every peeping eye see it working. Even the means by which Leibniz *wrote* his discovery surpassed Newton's in terms of accessibility, and it's noteworthy that the calculus of today uses Leibniz's symbols, which are principally "Roman" letters, like $x$ and $y$, with symbols like $\delta$ for delta function and $\infty$ for infinity. But though we start with a definite and specific feud, the lines become far more blurry—steeped in venom and anger and even hatred—as the feud progressed. "When lions battle," Newton wrote to Leibniz as an opening shot, "the jackals flee."[41] Whatever else the feud represents, it returned, ultimately, to God and to the faith-based order of the clockwork cosmos at large.

What would it mean to suffer a clockwork world? In 2013, I paid a visit to the world's oldest, largest still-functioning astrological clock in Prague, Czech Republic. Developed more than six hundred years earlier by clockmaker Mikuláš of Kada and Professor Jan Šindel, a teacher of mathematics and astronomy, the clock represents the universe in all its

---

* It's the more surprising because the two had been conversing by letter since 1677, and apparently recognized that they were on the same trail though by different methods. One explanation comes from Tasaday's examination of "lost" letters (2007). Supporters of both men engaged in public dispute, and before long, the titans were forced to bring their claims to the public. Newton, better connected and already a practiced bullfighter for his reputation, necessarily had the advantage.

perfection.[42] A calendar dial and automata appear in the design, including a skeleton with an hourglass and a bell. Enormous and imposing, the Gothic structure comes to life at the hours. While each figure has only a single action, the result of so many pieces moving at once—in addition to the astronomical dial itself—gives the impression that the entire clock may itself be one giant automaton. *Whitechapel Gods'* grim streets glow beneath the constant watching faces of clocks, each one a kind of messenger and reminder of Grandfather Clock, joyless automaton of logic. The Prague clock's imposing sense of scale hovers over the court, an unsmiling reminder of man's place in the world and his duty to the unseen but all-seeing Creator.* It leaves you with a feeling of your own smallness, but also of something brilliant, mechanical, and comprehensible: God made the earth, and the earth is the center of a beautiful machine. Except the clock also suggests that this perfect order continues along without intervention, that the motion—so ordered and perfect—needs no one to set it to rights. Newton never meant to promote a self-ordered world, but the seeds were always there, and Leibniz plucked them out in all their heretical potential. Originally in a letter to Princess Caroline of Ansbach in 1715, Leibniz claimed that Newton's work went against Christian doctrine. The whole matter came to public attention through the "Leibniz-Clarke" papers, an extended, long-distance debate between him and Samuel Clarke, friend and supporter of Newton. They've been called the most influential philosophical correspondence of all time, and while it never brought Leibniz the recognition he craved and deserved for his own part in the discovery, it brought to sudden wide attention the full potential of Newton's oversight—the first wrench in the gears.

---

* Other early automatons were less imposing, but no less fascinating—Leonardo da Vinci's sketchbooks contained the ideas for numerous mechanical structures, including a self-propelled cart, and the British Museum contains the mechanical ship built by Hans Schlottheim for the Kunstkammer of the Elector of Saxony in Dresden in about 1585. Its job was to announce dinner, rather than the hours; the galleon and other pieces like it were developed by clockmakers for wealthy or imperial patrons.

Leibniz begins by describing Newton's "odd opinion." According to his doctrine, he claims, the Almighty "needs to wind up his watch from time to time: otherwise it would cease to move. He had not, it seems, sufficient foresight to make it a perpetual motion. Nay, the machine of God's making is so imperfect, according to these gentlemen, that he is obliged to clean it now and then by an extraordinary concourse, and even to mend it, as a clockmaker mends his work; who must consequently be so much the more unskillful a workman, as he is oftener obliged to mend his work and set it to right."[43] Newton never meant a clockmaker God, but his work left room for precisely that construction—and it was, Leibniz quickly pointed out, "a very mean notion of the wisdom and power of God." If Newton's math allowed us to see how the machine worked, its constant need to be "righted" also suggested a world imperfect, constantly in need of attention. Newton's own assertion that the world would collapse without its God, ironically, left open a gap for speculating about its inverse, the self-winding machine. He had "inadvertently" given aid to skepticism, to doubt, and to a secular science that would not raise its head until the mechanistic philosophy of the eighteenth century. As Dolnick explains, in "bolstering" the cause of science, Newton "demoted" the agency and ability of God . . . and by extension, of man. If the universe were a machine, Judith Drake (a writer of educational polemics) complains in 1721, then humans themselves might be little more than clockwork.[44] In such a world, meaning had no meaning, morality no purpose. For all the good engineers of science might do in the future, the idea of a Godless clockwork world left no room for man's immortality. It meant chaos on earth would not be rewarded with the order of heaven. Newton fought back, accusing Leibniz of heresy instead, and the conflict continued in bitterness for the rest of their lives. Here again, fortune favored Newton, who, in his capacity as president of the Royal Society, ruined Leibniz's chances of ever claiming calculus for his own. Newton outlived his rival, and claimed valiantly that he "broke [Leibniz's] heart." Leibniz did his own irreparable damage to Newton, however, though Newton would not live to see it.

From the chaos, order. From order, chaos. When Peters depicts the final destruction of the *Whitechapel Gods*, he relies on metaphors that mix organic and mechanical processes in visceral and eviscerating poetry. The sense is symbolic; we know only that Oliver carries a device inside him and that he must plug into the machine itself: Oliver's thoughts "ended" and his "mental space became home to the calculations of the machine." To begin with, all is order and harmony. Then, "a black shape appeared to mar the infinite perfection. The ticks and tolls beat against it, but it refused to be made compliant. The shape rushed out, slipping into the empty spaces between Grandfather Clock's thoughts, and began to devour everything it touched."[45] Grandfather Clock falls out of rhythm. "His million sounds phased away from one another, changing pitch and timbre," writes Peters, and "the Great Machine collapsed from within."[46] Newton's God doesn't perish in quite so surreal or so speedy a fashion, but his sense that only the Almighty held the world together, that comets and other cosmic messengers were sent to reenergize the clockwork, that numbers—like the alchemy of his youth—must be protected from the unworthy, a magical language only he and his God could understand. But something surprising comes of this, a substitution that even steampunk fiction and sci-fi dreams could not have predicted.

In 1735, Alexander Pope penned an epitaph in honor of the man, in which he lifted him to the stature of a god: "Nature and Nature's Law lay hid in Night:/GOD said, *Let Newton be!* And all was Light." Newton had become, by 1691, the champion of religious skeptics and secularists, his views offering an alternative to faith in the creator that would become the hallmark of Deists (and, notably, "founding fathers" like Benjamin Franklin, who believed that God had wound up the universe and then walked off stage). Wootton marks the end of the seventeenth and the dawn of the eighteenth century as "the disenchantment of the world"; David Hume's *An Enquiry Concerning Human Understanding* of 1748 and other works picked up the theme, offering the clockwork universe as a possibility that sounded the death knell of witches and goblins, but also of belief in miracles and then, at last, even in God him- (or her-) self. Unlike Leibniz, buried in an unmarked grave without friends, Newton's remains lie at Westminster Abbey, where men still pay him homage. But though Newton sought to prove God a mathematician, in the end, Newton himself became the God of Mathematics. *Man* would replace *God* in the ensuing scientific age. And that meant grappling with entirely new questions about our place in the world, in life, and after death.

Frederick Nietzsche warned that those who seek to destroy monsters should be careful to avoid becoming monsters themselves. He warned, too, that if you look too long into the void, it may very well look back. In the latter half of his life, after publishing his *Principia Mathematica*, Newton turned his back on chemistry in order to dwell in the dark mysteries of alchemy, claiming that "All things were created from One Chaos by the design of One God [. . .] so our work brings forth the beginning out of black chaos."[47] And yet, he also authored the great proofs on which our modern science has been built, not just calculus but the laws of motion, gravity, and measurement. Not before and possibly not since has a single human being achieved so much, with so little, in so short a time. In seeking to discover divine order, however, Newton and his contemporaries introduced the age of skepticism. If God is dead, are we left with only dread tech? And what happens if the machine of the world breaks down in the end, leaving nothing permanent? No wonder the inventors who come after Newton find themselves deeply committed to powering the future. Every invention, Elting Morison reminds us, is a *dislocation*. Every new thing replaces an old thing. The message of our past creeps into our fictions as well as our facts, our stories as well as our histories. From the backlash and backbiting of Leibniz and Newton to the cautionary tales of latter-day thinkers and writers as disparate as Matthew Arnold and Mary Shelley and Jules Verne, the message is the same: "what is supposed to be humane about humanity gets ground up into smaller pieces" as the machinery clanks ahead.[48] In the next chapter, the sacred order doesn't concern heavenly bodies, but human ones, and controlling the chaos will require more than math; it wants a brand new kind of *machine*.

## TWO

# Clockwork Boy and the Mother Machine

I [. . .] took a voyage to visit my mother Nature, by whose advice, together with the help of Dr. Diligence, I at last obtained my desire; and, being warned by Mr. Honesty, a stranger in our days, to publish it to the world, I have done it." So begins the preface of a most curious work, the 1652 *The English Physitian*, by Nicholas Culpeper, an astronomer-botanist who promises to reveal the heavens in the human.[1] To Culpeper, the body and all its wondrous anatomy were not divorced from nature, nor from the cosmos, but rather, the body was a "Microcosm," a mirror of the cosmos in miniature, a flesh-wrapped star-map with which the heavens had intimate dealings. Diseases, he claimed, varied with the planets. Likewise, every plant in nature corresponded with a planet or planetary alignment. The diseases of Jupiter could be cured by the herbs of Mercury, ailments of the moon by the plants of Saturn, and those of Mars with vegetable Venuses. Culpeper finds himself in company with Kepler and Newton (rough contemporaries); *"Because out of thy thoughts God should not pass, His image stamped on every grass."*[2] God built the world, and so the body, and a good physician could "read" these hidden signs in flesh and blood as an astronomer reads stars. "No man ought to commit his life into the hands of that Physician, who is ignorant of Astrologic," Culpeper claims, because such "is a Physician of no value."[3] Kepler and Newton and Leibniz looked into the heavens and saw a glorious mechanism, the orderly machine of an Almighty mathematician. Culpeper's little book, bound in leather, block printed, and comfortable in the hand of the traveling apothecary, suggests a cosmic order to the body—*cosmic* but still *organic*. Culpeper's work (and others like it) got all the facts wrong, but offered succor in its ardent belief about God's plan for his creation. You are a world unto yourself, they claimed, specialized and cosmic and made of star dust, a child of Mother Nature, your every limb as moistened by her dew as the grass of the field. But times, they were a-changing.

In 1664, the same year that Henry Power pronounced the dawn of "new philosophy" and just before Newton launched his magical thinking, France's most renowned philosopher turned the focus from the celestial and natural bodies to *mechanical* ones. René Descartes's *Treatise on Man* (*L'Homme*), published posthumously, represents the culmination of his work on physiology and mechanistic psychology. In the years after its publication, scientific and medical thinkers alike would begin to see the messy, complicated, mysterious chaos of our ephemeral "heavenly" body reduced and refined as "a fine-tuned machine."[4] Like the stars above, or like the planets on their revolutions around the sun, the body might be atomized, rendered accessible, controlled, and contained. And like the complicated brass wheels of a revolving planetarium, the body might also be modeled . . . and even replicated. "If life was material, then matter was alive," argues historian Jessica Riskin, "to see living creatures as machines was also to vivify [or bring to life] machinery."[5] By comparing the body to clockwork and hydraulics, did Descartes begin an unintentional revolution, not only challenging *what it means to be human*, but *what it means to be a machine?*

## *The Human Machinery*

I was four years old when I saw my first automatons. The "Swiss Chalet" (Sugarcreek, Ohio) boasted a carved roofline of snow-capped pines, strange in the gritty heat of July, plus wooden lattice and a working waterwheel. It's the sort of roadside attraction that summer drives are made for: the world's largest cuckoo clock. A peaked gable housed a single window, under which a clock told the hours. I spent an eternity of minutes, my eyes fastened on the shutters, waiting for the gong to sound. *Strike, strike, strike*, and the panes sprang open for the cuckoo to sing. I remember fascination: a clockwork bird, open beaked, ducking in and out of its hole in the machine. That was but the beginning. Following the herald, five mechanical players bumped and banged their way onto the stage. With painted faces stuck at smiles, a fully automated polka band stuttered out. Each stood nearly as tall as me, and each turned a wooden head in my direction, unseeing and unreadable. For reasons I couldn't articulate then, the clock haunted me. Its bright colors and the too-loud music, the bird, the band: they danced on in my nightmares with their eerie not-alive life. I'd just experienced something important, feelings of linked fascination and dread that accompany our long history with technological innovation.

The lever, the pulley, and the screw: in many respects, these were humanity's first "machines," tools that could ease the labor of the body when moving a load.[6] Galileo published *Mechanics* in 1600, the most advanced of his day—but even he considered machinery in narrow ways, as weights and measures and forces. Galileo believed in "atomism," or the breaking down of something larger into smaller and smaller parts that dodge around in empty space (it's where we get our word for atoms). This theory, based on the work of ancient Greek scholars like Democritus and Epicurus, makes up one part of mechanistic philosophy. The other half comes from Descartes's peculiar idea about "corpuscular" theory. At its most basic, corpuscular mechanics just means that instead of empty space, you have a sort of fluid, meaning that invisible forces touched and interacted with each other in a continuous substance. (Newton had yet to discovery gravity, remember.) But in following his ideas, Descartes faced a series of seeming anomalies—one thing must interact with another thing. The interaction must be a force. But the force couldn't be seen. Was it right to believe in what you could not see? Was it right to reason from the invisible? In his questioning, we glimpse the still-flickering image of a brand new idea Descartes would begin to explore in *Le Monde* (The World) and make more concrete when he wrote *L'Homme* (Man). Culpeper began by mapping human bodies with heavenly ones, and he never once doubted that he read the signs correctly. Descartes begins by doubting not only the heavens above, but the earth below, all he can see, and all he can know. In the process, he gets from the *God* to *man*, and from *man* to *machine*. And he invents one of the most important aspects of scientific method in the process.

In 1619, René Descartes was a young soldier, overwintering in Ulm, a Bavarian town on the banks of the Danube. Frigid, bitter wind howled around the housetops and down chimneys, and Descartes huddled next to a wall stove for warmth. With little else to occupy him, he read philosophical treatises by candlelight, only to discover that most all of them contradicted each other. So continued the long dark days, and the longer, darker, nights—and all the while, Descartes read and thought, thought and read. What was to be believed if the greatest minds could not even agree? In that troubled state, alone and without friends, he began to doubt, and not a particular aspect of science. All the ragged bits and pieces stuck on or struck off over time, all the theories and ideas, every experiment of thought and every innovation seemed equally untrustworthy, equally flawed, both unproven and unprovable. Chaos, in other words. A near hopeless idea struck him: could a single person start over from scratch, working only from his own observations? Drowsing and low-spirited, he fell asleep—and had three dreams. The first was a whirlwind of phantoms and far-off lands, the second all sparks and thunderclaps, and in the third, strange beings bid him read from a book: "*what path shall I take in life?*" Descartes awoke with a new direction for his life: *he* would be the one to start over, and build knowledge from the ground up. "It often happens," Descartes explains in *Discourse on Method*, * "that a private individual takes down his own [house] with the view of erecting it anew [. . .] when the foundations are insecure."[7] We should recognize again the sense that something must be destroyed before anything can be built, but Descartes doesn't want to begin even with foundations. Begin, he suggests, with nothing. We cannot accept what we have not come to know for ourselves, but even "our senses sometimes deceive us." If even the eye cannot be trusted, Descartes concluded, then "all the objects that had ever entered into my mind when awake, had in them no more truth than the illusions of my dreams."[8]

Most of us dream. Sometimes a dream seems so detailed, so "real" that in waking, we can't determine happenings from illusions. Doesn't the dream-world feel real, and even excite all the emotions of life from fear to excitement? Descartes's journey begins with a dream, and his next thought experiment in the dark Ulm winter was to ask: How do I know that I'm real, that I am the dreamer and not the dreamed? Descartes answers the question with his now famous line: *Cogito ergo sum*, or, I think, therefore I am. To doubt requires a doubter; for Descartes, at the core of all things, there must be a thinking, questioning, doubting being. From this, he forms the first principle of his method, a system of discovering knowledge that we take for granted today: never accept anything for true on someone else's

---

* The preface of *Discourse on Method* lays out his (incredibly ambitious) plan in six parts: in the first, "various considerations touching the Sciences"; in the second, the "principal rules of the Method," in the third "rules of Morals" from the Method; the fourth he dedicated to the existential problem, the existence of God and the soul; and the fifth investigates the body itself, the motion of the heart and how body and mind interact (or not). In other words, his first five sections cover no less than all the known sciences, investigation methods, morality, religion, and medicine—and in the fifth, he at last gives "the reasons that have induced him to write" (without hope of profit or gain, but only the advancement of science).

word alone, avoid all prejudice, and believe nothing unless "presented to [the] mind so clearly and distinctly as to exclude all ground of doubt."[9] Descartes reverses the maxim of the ancients. The point wasn't to *accept*, but *prove*, moving "little and little" and "step by step," from things of least complexity to things of greater complexity. We can put it in perspective if we think of Kepler's soccer-ball universe; he assumed a grand design, enormous in complexity, and then tried to work backward, forcing nature to divulge her secrets. If you begin committed to an idea, you may bend the facts to fit them. Descartes assumed only that he could think, and proceeded from this first kernel outward. By a circuitous route, proving one thing after another, he arrived back at a belief in a Perfected Being—a God. But it also led him to conceive of human bodies in a brand new way.

"I have now explained [. . .] from the very arrangement of the parts, which may be observed by the eye alone," concludes in Part V of *L'Homme*, that the body works via *mechanics*, "as does the motion of a clock" and "its counterweights and wheels."[10] He compared the body to hydraulics, too, that whirr and bubble like fountains in a courtyard. Each part of our organism connected to the whole and—as gears and cams—made up a great and wonderful machine. But there arose a problem. On the one hand, Descartes was far more willing than Newton to suppose a clockmaker God, a "Grandfather Clock" that wound up the universe and then walked away from it. Though he never in fact compares the universe as a whole to a man-made machine, for Descartes, the universe self-assembled, operating as "a self-fabricating automaton."[11] How different from Newton! The clockwork didn't need winding. But did that leave room for the mind, which was "immaterial"? He solved the problem by thinking of the mind a bit like he thought of corpuscles, where the soul, like those invisible forces, acted *upon* the body without being *part of* the body.* The force which moved us was,

---

\* A simplified version of what in fact is a highly complex argument; for a deeper understanding of Descartes's scientific and philosophical vision of soul/mind/body, see Desmond Clarke's *Descartes's Theory of Mind*.

to quote philosopher Gilbert Ryle, "a ghost in the machine."[12] All bodily activity was material and direct, mechanical, and even "greasy" (as Descartes himself admitted). The mind, however, was free from this arrangement, uncoupled from and untroubled by the body. The philosophy that rises from Descartes's method borrows his name—*Cartesian Dualism*—but in fact diverges from his work considerably. Just as Newton's insistence on a well-ordered world begged the question about the existence of God, the mechanical understanding of our bodies left a gaping hole of doubt about the human soul. His work had been to reconcile his scientific pursuits with theological views. The materialists that came after him felt no need to do so. What was man, they argued, but the perfect automaton?

## Clockwork Boys

*"You know, machines never have any extra parts. They have the exact number and type of parts they need."*
—Brian Selznick, *The Invention of Hugo Cabret*

In April 2015, Toshiba unveiled ChihiraAico, a robot built to resemble a young Japanese woman.[13] She talks, sings, gestures, and even *cries* using a responsive artificial intelligence matrix that reportedly "disconcerts" those who interact with her. A year later, at the March 2016 South by Southwest exposition, Hanson Robotics introduced their own "female" humanoid, Sophia, possessed of sixty-two facial expressions, including deep sadness. To make the robots more lifelike, some designers use casts of real-life models, right down to the teeth. These unusual creations strike visitors with the same unsteady feelings that followed my encounter with the cuckoo clock so many years ago. Why do we make such machines? And why give them emotive expressions that they cannot possibly feel as we do? Miraikan—Japan's National Museum of Emerging Science and Innovation—showcased an exhibit titled "Android: What Is Human?" Exhibit designer Hiroshi Ishiguro argues that these are the wrong questions. In Japanese culture, says Ishiguro, "Everything has a soul. We don't distinguish between humans and others." The desk, the chair, the microphone into which he speaks—and the complex automatons he creates all potentially have "souls."[14] And because of this, he claims, they can accept this "new type of creature" easily. Here, the difference breaks down between the man and the machine.[15] Materialists of the seventeenth and eighteenth centuries likewise insisted that mind and soul were dependent entirely upon the "physical properties of matter." If man is a thinking being, and machinery is responsible for thought, you don't need a dirigible to get from "humans may be machines" to "machines may be human."

One of the first to make that jump was a man named Julien Offray de La Mettrie. Taking Descartes's *L'Homme* as a point of departure, his book *L'Homme-machine* (or man-machine) broke new ground in 1748.[16] Thomas Willis came along shortly after with ideas about brains and nerves, a leap, says Allison Muri in *Enlightenment Cyborg*, from strict clockwork to "feedback engine."[17] It was also Willis who coined the term "cybernetic," and in fact, he's the first to suggest that the mind was a communications hub. In other words, the human is a "sensible machine," communicating complex thoughts through tissues and nerve endings, the cogs and wheels of the human body.[18] By the mid–eighteenth century, at least, humans were being described as tidy, well-ordered machines. We could look into the complexity of gearwork (said philosophers and even physicians) as into the intricate web of our being; here were the origins of thought and of purpose. Here, and not in the mythologized, incorporeal, ungraspable soul. So the question for those who followed Descartes was not "Do machines have souls?" but rather "Does man have one?" Materialist thinker Baron d'Holbach flatly denied any soul whatever; his *System of Nature* (1770) begins by asserting that man would be far happier to abandon the idea altogether and embrace his mechanical nature.[19] And yet, as apologist George Berkeley writes, if man is "but a piece of Clockwork of Machine: and that Thought or Reason are the same thing as the Impulse of one Ball against another," then he cannot be held accountable.[20] He becomes not the wondrous machine but, as Judith Drake had suggested, a *mere machine*, "a sort of Clock-Work, that act only by the Force of nice unseen Springs, without Sensation, and cry out without feeling Pain, eat without Hunger, drink without Thirst, fawn upon their Keepers without seeing 'em."[21] In the wake of Descartes's dreams, we don't get a unity of science, but instead the earliest beginnings of two very different stories: in one, the valorization of human machinery, like Hugo's wistful assertion in Selznick's novel that all machines have precise parts, working precisely—that hopeful assumption that "if the entire world is a big machine, I have to be here for some reason."[22] *The Invention of Hugo Cabret* principally concerns a writing machine, an automaton that accomplishes by those very "nice, unseen springs" the task of careful, exact, written script—and the key to the novel as a whole. Parts for a purpose, intelligent design at its finest. The seventeenth and eighteenth centuries would witness the building of clockwork bodies by the score, ordering the internal and setting it to rights within the grand system of an automaton universe. But there is a second story, and it stretches a shadow over the first, like long fingers of smoke that besmirch the gears. Its representative and herald stands as evidence of something far more unsettling—and she, like Mama Engine, is also a mother.

It began with Nicholas Culpeper and his consideration of bodies, heavenly and earthly, in medicine. In 1646, his contemporary Athanasius Kircher published a similarly cosmic work, *Ars Magna Lucis et Umbrae*. Compared to Leonardo da Vinci because of his vast publications and an incredible understanding of everything from geology to astronomy and theology, Kircher worked to provide evidence of the cosmic relations between the human and the divine. His diagram matches complexity with artistic unity, a kind of sublime chart that lends credence to Culpeper's cosmic botany

[Fig. 2]. That both men had it wrong isn't entirely the point. It's that they may have been as right, or at least as right as anyone else at the time. Who would contradict? Medical knowledge had not drastically improved in *1,400 years*. Until as late as 1543 and the publication of Andreas Vesalius's *Fabric of the Human Body*, medical doctors relied almost entirely on the anatomical work of Galen, though published around 100 C.E. Anatomy training didn't really amount to much in the centuries before Vesalius; a professor (who had probably never performed a dissection) would stand at a raised chair and comment on the work of Galen, turning the pages and describing organs while his assistants would actually take apart the cadaver in front of a theater of students. Galen himself had never dissected a human either. This had not been allowed in his day and remained taboo for centuries after—and humoral theory, that we are comprised of four different humors (black bile, yellow bile, blood, and phlegm), held sway well into the nineteenth century. The most common reason for using a surgeon's knife on a living body was for bloodletting, a practice supposed to cure everything from infection to mental distress. Called "barber surgeons" because they might just as likely shave your beard or cut your hair as operate, these men (and all were men) understood nothing of germs or cleanliness, still less of the body's careful organization. Physicians fared somewhat better, in theory, though recipes for medicines sound like alchemical preparations; rabbit dung and rosemary for an aching tooth, roasted raven's heart to cure "fits."

The preparations for Culpeper's treatise sound benign by comparison, mostly plants to fortify the body, tissues, and blood. In fact, though all of Culpeper's contemporaries understood the power of blood as a life force, it would take William Harvey, through a long course of dissection on dogs, to prove that it *circulated* through the body. Until 1628, physicians assumed the body made all the blood each day, and different sorts of blood for veins vs. the arteries. You can imagine how many patients bled to death from their various treatments when doctors so overestimated our ability to re-create those precious red and white cells. When Charles II suffered a stroke in 1685, doctors drained him of at least two cups just to start, then forced him to undergo an enema and a purgative, all means of getting the ill humors out of the body. Failing that, they rubbed him with pigeon dung and powdered pearls, and applied hot irons to his feet. As a last resort, they even practiced a form of *corpse medicine* (dead bodies were consumed in powders for centuries in multiple cultures)—extract of a human skull.[23] It did not work, though even Robert Boyle, member of the Royal Society, father of modern chemistry, and pioneer of the scientific method was a keen supporter of corpse medicine. For over two centuries in early modern Europe, says Renaissance historian Richard Sugg, "the rich and the poor, the educated and the illiterate all participated in cannibalism on a more or less routine basis."[24] The Age of Enlightenment knocked on the door of the nineteenth century before surgery was anything but a death sentence. We might be machines, but there could be no assurance that the gears could be put right once they had broken down.

Novelist Fanny Burney recounts her mastectomy in 1811: "Yet—when the dreadful steel was plunged into the breast—cutting through veins—arteries—flesh—nerves [. . .] I began a scream that lasted unintermittingly during the whole time of the incision." Awake, she hears, as well as feels, the grating of the blade, and:

> When the wound was made, and the instrument was withdrawn, the pain seemed undiminished, for the air that suddenly rushed into those delicate parts felt like a mass of minute but sharp and forked poniards, that were tearing the edges of the wound—but when again I felt the instrument—describing a curve—cutting against the grain, if I may so say, while the flesh resisted in a manner so forcible as to oppose and tire the hand of the operator, who was forced to change from the right to the left—then, indeed, I thought I must have expired.[25]

S. M. Peters's novel opens with a surgery too. "Bailey was not surprised when the doctor's first incision drew up something darker than blood," explains the narrator. "The patient writhed and struggled in the bed, fighting a pain that distorted his features into something less than human."[26] "Clacks" infects this steampunk dystopia, not so different from cancer. Reliable forms of anesthetic weren't available until at least the middle of the nineteenth century, and even then, they were not always safe or precise. Worse, germ theory didn't catch on for decades after discovery; Joseph Lister had to defend his ideas about antiseptics against vicious critics and skeptics well into the 1870s (something I'll return to in chapter 8). The surgery of mid-Victorian England, in other words, didn't look very different from those at its start, but as terrible as anesthetic-free mastectomy sounds to modern ears, Burney survived. The brutal bludgeoning of the knife blade, the wretched sawing motion, the unfathomable pain are nonetheless *enlightened* medicine—far better than the mad guesswork of body star maps or strange potions of supposed healers, because while ingesting corpses may not save you, *studying* one could.

Vesalius ushered in a new means of practicing and teaching anatomy. Under his tutelage, and through his enormously powerful anatomical atlas, generations of doctors finally learned what went on beneath the skin . . . but is it any wonder that this heaving, wriggling, discolored and dis-proportionate cluster of organs and meat remained so long a mystery? To understand the whole, we must investigate the constituent parts. Newton explained God to man; Descartes intended to explain man to man. The mechanical metaphors of *L'Homme* intended to render bodies plain and legible. Descartes himself built automatons—including one that he named for his lost daughter—but the true test of man-machines would come *after* his death, and from the hands of makers. What better way to understand the body than to build one?

A young boy sits at his writing desk. He wears a red velvet coat and lace cravat, his bright eyes moving back and forth under sleek black curls. His head turns, stiff, but graceful, and fingers dips a quill into a nearby inkpot. I've used a quill. Too much ink, not enough, the end is to sharp or too blunt, and it spatters. But he doesn't spill a drop. Built by Swiss clockmaker Pierre

Jaquet-Droz, *The Writer* debuted in 1774, two years before the American Revolution, some sixty years before the dawn of the Victorian age. The serious-faced "boy" arrived during a flurry of clockwork inventions as one of the finest automatons of his day. Today, the automaton resides at the Musée d'Art et d'Histoire of Neuchâtel, Switzerland [Fig. 3]. But despite being more than 240 years old, his movements remain precise and exact—and a joy to watch. Jaquet-Droz could never have imagined the tricks of laser precision and computer modeling. Even so, and operating under far less favored conditions, the clockmaker miniaturized 6,000 working parts to fit inside the twenty-eight-inch-high automaton [Fig. 4]. All the automaton needs to function, from parts to power source, were neatly puzzle-fitted into his little body: a self-contained, self-moving machine. Cam technology made it possible; the teeth, crafted small and stacked vertically, transferred linear motion to lateral movement. Sliding cam-followers translated those same grooves—through the goose quill—into *words*. All this makes the Writer a mechanical marvel, but Jaquet-Droz's greatest contribution wouldn't be entirely understood for centuries. He inserted a wheel of letters and other characters that could be refitted and replaced. Move the letters, change the message; the Writer could be *programmed*.

*The Invention of Hugo Cabret* depends on the same remarkable ability: the plot revolves around a mysterious automaton and the preprogrammed message he transmits. But Jaquet-Droz's second mechanical wonder, the harpsichord player, not only played her instrument, she also moved with gestures and looks that "expressed delicate feelings"—that is, with emotions.[27] Adelheid Voskuhl, author of *Androids in the Enlightenment*, explains that by nodding, moving her head to the rhythm, and even sighing, the automaton communicated *her humanity* to the audience.[28] Other inventors went further still. Jacques de Vaucanson, for instance, introduced a flute player that actually *breathed* into his instrument, mouthing the flute with flexible lips and tongue.[29] Vaucanson planned to build a machine with a working circulation system too, but couldn't get enough India rubber for the parts (he settled instead of building the "defecating Duck").[30] In all of these creations, we see a strange commitment not only to looking like a body, but functioning like one. The machines drew excited crowds; they certainly qualified as *spectacle* as much as science, but here we have science all the same. Our internal functions may no longer be a mystery—the lumps of soft tissues and muscles, pumps and signals working undercover and in utter darkness have at last been laid bare. With anatomy, the light had dawned, but it was only with machines that those in the seventeenth century could put this new understanding of our well-ordered parts into practice.

Automata, says Voskuhl, "are often taken to be forerunners and figure-heads of the modern, industrial machine age," a time when everything from the economic to the social to the biological became "mechanized."[31] We know, today, there is a cost to such meddling. If we recall Elting Morison's questions—Who changes? Who gets hurt? Who profits?—we must reckon with the bits of humanity lost in the process. The harpsichord player sighs, as "though her heart would break," but is affecting emotion the same thing as feeling? Descartes struggled against the notion, but inventors who came after largely dispensed with his "messy" philosophy about the mind and body. Or to put it another way, everything became body, all of it part of the works. The period between the 1730s and the 1790s, writes Jessica Riskin, was one of "simulation," in which "mechanicians tried earnestly to collapse the gap between animate and artificial machinery."[32] That included building strange speaking devices, essentially floating heads with flexible mouths and artificial lungs, as well as the bits and pieces of torso or limb described by Vaucanson. Man and machine were so linked in these devices that a surgeon helped Jaquet-Droz build the writing automaton's skeleton;[33] likewise, though reversed, the best maker of steel trusses for ruptured human bodies was a surgeon who also worked as a watchmaker. William Blakey could construct personalized trusses for everything from hernia to prolapsed uterus using steel, ingenuity, and watch springs.[34] The crossover of body and machine becomes increasingly tangled, and the title of La Mettrie's very controversial—and very popular—*L'homme Machine* says it most clearly: *Man a Machine*. Regarded as a "radial materialist," La

Mettrie might be seen as the culmination or fruition of Descartes's ideas, while entirely removing the concept of "soul." The brain, he explains, is only "a well-enlightened machine" and the body "an immense clock." Prefiguring the science fiction cyborg, La Mettrie introduces "the human body [as] a self-winding machine, a living representation of perpetual motion." This biological automaton doesn't radically differ from those invented by the clockmaker; his workings are all encapsulated within the living frame, just as the writing boy's gears fit snug into his body. The animating principle resides "in the very substance of the parts [. . .] in short, the whole organization of the body."[35] In their work on artificial intelligence, Stefano Franchi and Güven Güzeldere point out that this way of conceiving the human being led to the mechanization of intelligence in the modern era.[36] In other words, without Descartes and La Mettrie, we could not have a designer like Hiroshi Ishiguro or a robot like Toshiba's ChihiraAico. But philosophy does not, by itself, manufacture the future.

Elting Morison claimed that whether or not there are such things as true inventions, "it is clear there are inventors, or at least there is a syndrome, as clearly defined as any neurosis, possessed by [those] who are said to invent."[37] And we can find at least two defining features belonging to everyone I've mentioned here, and also to Kepler, Newton, and Leibniz in chapter 1. First, though their reasons varied, each found himself unhappy with or opposed to the condition of things as he found it. For Descartes, the very system upon which knowledge had been based seemed little more than unsteady shale, crumbling away underfoot. He would find a new and better foundation. (And in fact, Newton goes on to finish what he started, almost literally, when he establishes his "laws.") For Jaquet-Droz and William Blakey, the problems were localized: as makers, they spent their time investigating tools and methods and styles. Using the precision of the clockwork, they had a grand object in mind: a more perfect machine. That it had not been done before made no difference; they would do it. Jaquet-Droz created the first programmable machine—and Blakey turned to the latest in steel manufacture to design newly personalized trusses, so superior that his clients flocked from all over Europe. But the popularity of either "machinery" could never have come to be without the secondary trait of invention: the ability to, even the obsession with, spectacle, salesmanship—the *show*. Descartes, like Newton, claimed he was the chosen one; also like Newton, this story is "curated." Neither man actually had a lightning strike of inspiration in quite the way they describe, and their journals and letters prove that events happened out of order, over time, and often with the aid of intelligent conversation with others. But inventing the modern world by committee doesn't have the same flair, does it?

Descartes fought an engineer named Isaac Beeckman over his ideas of "physio-mathematics" (a general science of all mathematics including physics and alchemy) and music. Beeckman claims in 1630 that he communicated certain scientific problems to Descartes for his *Compendium Musicae* on the mathematics of music—and is accused of plagiarism for his trouble Like Newton and Galileo, Descartes refused to acknowledge his debt to the other man because, as Wootton puts it, "he had told himself that he was building a new philosophy single-handedly" and any dependence upon another "was intolerable to him."[38] The stage is a powerful tool, and the inventor does not wish to share it. The clockmakers and truss manufacturers had perhaps more practical reasons for not wanting to share the limelight; they competed quite literally for paying customers. Blakey's indomitable wife helped to solve the problem for him: access to and communication with steelmakers and with the French court. As a "privileged merchant" she became the exclusive seller of steel in the nation.[39] In applying spring-steel properties from watchmaking to trusses, Blakey secured a "privilege," too, not unlike a patent. Establishing exclusivity allowed the Blakeys to advertise widely while cornering the market; they applied their advertisements to newspapers, leaflets, and even mailers.[40] Vaucanson even evaluated Blakey's instruments for the Académie des Sciences in Paris where he debuted his own automatons—the flute player, a galoubet (pipe) player, and the duck—in 1738. Excited crowds would gather, educated and otherwise, to see, to cheer, and even (in the case of a hoax called the Chess Player) to faint at how lifelike a machine might be. La Mettrie uses Vaucanson's automatons as his examples par excellence, but by comparing humans to soulless animal machinery, he drew his own crowds through controversy (in which, it seems, he delighted). Disapproval or disdain of the usual course, plus a maddening desire to be the "chosen one," to be first, to be best, to be *only*—and an equal ability to operate as stage manager, carnival barker, and salesmen: these inventors cast themselves as the heroes of an unfolding story. And together, they present the human machine and its well-ordered parts like the brass fittings and bright gears of the clockwork boy. "Each part contains its own more or less vigorous springs, according to the extent to which it needs them," concludes La Mettrie, a historical antecedent to *The Invention of Hugo Cabret*. What could possibly go wrong?

## The Mother Machine

A man stands in the center of the medical theater, surrounded by eager students. It's the mid-eighteenth century and he's delivering a baby. He fishes a spoon-like contraption on either side of the infant's cresting head and then scissors them together for traction; these are the newly introduced *forceps*. A tug, and the baby slides forth from the womb to great applause. It's an obvious triumph of new technology, but the doctor doesn't press the infant into the waiting arms of its mother. In all likelihood, he'd spend the next few minutes *shoving it back inside the womb*. It may look like a flesh-and-blood woman, it may even be wearing a petticoat and stockings, but the figure on the table is actually a contraption, a *machine*. Following on the heels of Descartes's hydraulic imagery and new dissection charts of the gravid (or heavily pregnant) uterus compiled by Drs. William Smellie and William Hunter, a new class of "Man-midwives" gives rise to a brand new idea. If the body was machinery, then surely reproduction itself, and the final breach of a human being into the world, might also be reduced to "clockwork."[41] As I say in an article for *Feminist Formations*, medical practitioners had begun meddling with the body's hidden and chaotic interior to control "female fecundity" (or fertility), shake off the horror and mystery of childbirth, and make the entire process a workmanlike affair.[42] But to do it, they would need to understand the entire female economy and peep in at processes never before glimpsed (and certainly not experienced) by men. As critic Bonnie Blackwell claims, man-midwives with engineering ingenuity would go on to "manufacture the kind of woman they could not find in the world"[43]—a mechanical mother.

I first came across this strange device in old medical advertisements. I'd been researching mechanical *habits*, behaviors learned by rote and practiced without thinking. In late seventeenth- and early eighteenth-century Britain, women (particularly those of middle to upper rank) were expected to perform two different, though related, acts of "reproduction." They were the begetters of sons, of heirs, of new British subjects. And yet, women were also responsible for the early education of their children. The dual emphasis of women's primary (procreative) and secondary (educational) function overshadows any other duties, but they conflicted—at least in the minds of male physicians and educators. As vessels for bearing children, women need not be educated at all, they argued; in fact, some medical treatises suggested that all the blood would go to a woman's brain and deprive her vital reproductive organs, rendering her sterile! Mary Astell, a seventeenth-century rhetorician, argues that men already thought of women as "machines, condemn'd every day to repeat the impertinencies of the day before."[44] To create a reproduction machine, they need only add the cogs and wheels and springs.

Consider: a woman at the turn of the eighteenth century might give birth twelve times in her life. Twelve times she faced the possibility of torn sphincter muscles, fistula (a gaping wound between the vaginal wall and the rectum or bladder), ruptured organs, or prolapsed uterus, where the uterus would hang down between the legs. It should be no surprise that William Blakey made a great deal of trade with damaged mothers, and delicate letters survive that show the embarrassed, painful state such women were left to live with. But despite these terrors, or maybe because of them, young women were shown pregnant wax dolls before marriage to prepare them for their destiny, and to be non-procreative remained the deepest of shames. But even with improved anatomies and models, most people had no idea how the woman's body actually functioned. Half of London was ready to believe Mary Tofts, a woman from Godalming, who claimed to have given birth to rabbits in 1726, and the papers carried stories of monster births well into the nineteenth century—even recycling bits from the 1500s and the star-crossed damnations of Tycho Brahe's nova. What lay in the writhing, roiling deeps of the womb? What procreative powers—and even what dread? Male doctors, bent on taking over midwifery from women practitioners, would take control of this mystery by fulfilling the worst fears of intellectual women like Mary Astell and Judith Drake. Just as La Mettrie dispensed with the second half of Descartes's mechanics by removing the concept of soul (and so collapsing the distinction between body and mind), the *mother machine* reduced a birthing woman to the disposition of her organs.

For centuries, the womb had almost a life of its own—the Greeks thought it could "wander," could get away from its usual housing and, in want of a baby, cause incredible havoc with a woman's health (this is, in fact, the root of "hysteria"—and also of *hysterectomy*). Remedies that now seem ludicrous were applied by healers, such as burning incense between the legs to coax the errant womb back into place. In one sense, the inventions of male midwives improve upon these mythologies by recognizing the actual structures and movements of the body. At the same time, rendering the female interior readable also meant simplifying it, and most of the early attempts were admittedly crude. It was with some suspicion, then, that I read the advertisements of Dr. William Smellie, a man-midwife of Lanark, Scotland, practicing in London in the 1730s. Come witness!—the papers, letters, and journals suggested—come see this "most curious machine," a "mock woman," a "celebrated Apparatus." The scant bits of information available tantalized in their details . . . She had a belly of leather, complete with a bag full of beer to simulate the bladder and other organs.[45] Her figure crisscrossed with ligaments, muscle, and skin to make it more lifelike; her supple form rested on real human bones, which gave her, in the words of eyewitnesses, the "Motion, Shape and Beauty of natural Bodies [. . .] with great Exactness."[46] Contemporary pamphlets describe the use of levers so that the machine's uterus contracted while in "labor," and the sale catalogue produced upon Smellie's death explained that she could be manipulated so as to demonstrate problems like malformation of the pelvic bones and their effect on birth. She wore clothes, petticoat, stockings, and shoes. And a surgeon, Peter Camper, claimed that the mannequin-machines were made with such remarkable skill that "*hardly any difference is to be noticed between these, and [features] in natural women*" (my italics).[47] Though, rather importantly, she didn't have a head. The female mind they considered dispensable . . . a horrific conclusion to the debate about Cartesian dualism that leaves us with a body and stumps, but no voice, no face, no self [Fig. 5].

The machine (or machines, if you count the machine's "babies") replaced the mother; but more than that, she's *an improvement upon* the mother in the eyes of would-be physicians. Smellie constructed elastic, reforming craniums for his doll fetuses . . . and once again, students refer to them as "more natural" than cadaver fetuses they had before practiced upon. Without flesh and blood, without the messy effusions of real biological beings, machines offered a comfortable simulation that did not *feel*. What a change from the clockwork musicians and their heartfelt sighs. I spent two years hunting this mad, strange device, following a trail that led from anatomy theaters to the close quarters of home delivery, from the pages of startling atlases to an archive lost to the Dublin fire, on two continents, in four countries. What I reconstructed defies classification but doesn't provide answers, so much as raise new questions.* Even so, in her mechanism, this mother machine

---

* Despite its popularity, the machine disappears by 1780. There are others, mannequins, plush works like the "phantom" (birth mannequin) of Madame du Coudray (the midwife of Louis XV of France), that remain. But the mother machine has gone, and though we've plenty of descriptions, I haven't found a single model, sketch, or student drawing of the device. Auctioned after Smellie's death to obstetrician William Hunter (physician to Queen Charlotte), and later sold to Dr. Foster, assistant master of the Dublin Rotunda, the machine simply vanishes.

carried the seed of an idea: machines do it better. As hauntingly explained by one of the onlookers, mechanisms like this laid "every material circumstance [of a woman's body] open to the naked Eye."[48] The device, hailed as a great success and engineering marvel, succeeded in revealing what lay hidden under curtains of flesh and blood and bone. This, more than any other technology of automation, far more than the clockwork boys and writing machines, shakes the foundation of what it means to be a body, a creative body, a female body. She helped train nine hundred man-midwives in ten years, she fascinated and horrified the public, and she vanished before the nineteenth century—but the mother machine changes the landscape of our greatest desires and most consummate dread. Boundaries do not elide; they bend and snap, and the consequences in fiction, but also in scientific and medical fact, are dire indeed.

In *Whitechapel Gods*, Grandfather Clock ticks along in perfect order, a body of logic and equations, like the writing boy with his unseen gears and perfect number of parts. But Mama Engine does not run like clockwork; hers is the filth and grinding noise of industry, the same shadowy, murky swill that Dickens describes in *Hard Times* or that can be pieced together from nonfiction reports like Henry Mayhew's *London Labour and the London Poor*. Mama's world does not tick and tock; organic as well as mechanized, Mama Engine's interior heaves up "a savage universe of pulsating desires given form in random and hideous shapes of iron, linked across distant leagues by strings of luminous, fiery coal."[49] She, just like Grandfather Clock, runs on "calculations," but these spin "like a flock of shining swallows, picking infinitesimal bits of data from the memories to link together and compile."[50] In other words, she works less like an algorithm and more like our own messy and importunate processes. More particularly, she is a *great mystery*, a creature with priests and rituals and servants and monsters. "The layers of her essence unfolded in sequence, revealing progressively deeper levels," honed at last to "an exacting point of fire, hungry to consume and smelt and mould."[51] Like the "wandering wombs" and their demonic births, Mama Engine spawns unknowables. This makes her power and her terror, and the reason why, from the start of the novel to the finish, characters on both sides attempt to control, contain, or destroy her. Mama Engine's own consort, Scared, takes a potion to enable him to couple with her. Now her master, he "reads" his bride "instant by instant with a knowledge as complete as God's."[52] The object is to *know*. The other simulation machines and automatons also created knowledge—but the comparisons remain positive, laudatory, exalted. In the seventeenth (and early eighteenth) century, says historian Minsoo Kang, the automatons and clockwork boys invited readers to appreciate "what a piece of mechanical work is man."[53] In contrast, the mother's "machinery" begs for intervention. A woman must be activated and "put right," her mysterious, messy, inefficient apparatus controlled and contained.[54] Instead of proving a mechanical order, the mother machine almost demonstrates the reverse; the body is *not* running like clockwork, but it can (and must) be *made* to do so. As doctors increasingly perceived "birth" as a scientific—or even pathological—affair, the woman's role either as the midwife or the mother shrinks, becomes secondary, subservient, even unnecessary. The machine, the forceps, and the doctor take her place; technology does the work. The mother has been erased, subsumed in the machinery.

Somewhere along the way, the worm turned; the very thing mankind meant to contain or control returned to sting at the heel. Many scientific, technological, and medical innovations begin exactly this way. Petroleum started as a replacement for whale oil, and has now driven the deadly wars of our age. Petroleum's derivative plastics make for excellent prosthetics and household goods, and yet pollute our oceans with particulate that lodges in the food chain and refuses to break down. But the horror of the replacement woman operated on a new level for her contemporaries, and especially for flesh-and-blood women. Elizabeth Nihell, a woman midwife who practiced at the same time as Smellie, complained that his "ingenious piece of machinery" missed the point: As much as it may looks human, she insisted, *it cannot feel pain*.[55] We return again to feeling, to the sensations that—more than our processing power—seem to separate the actual and virtual. Man-midwives trained upon something indistinguishable from the human body, but that could not cry out, could not call for help, and could not *die*. In short, by replacing the woman in labor with a machine all through the training, schools unleashed a generation of freshly minted doctors who may not be able to distinguish between feeling woman and unfeeling machine. The clockwork boy seemed to prove all was right with mechanism, that the human and the machine were ordered and orderly. If machines actually replaced humans, or if we manufactured (as Ishiguro has done) copies of ourselves in pristine laboratories, then perhaps some polite coexistence might come of our relationship with mechanism. But the mother machine doesn't reproduce her own kind; she gives birth to a dangerous overlap, a place where the machine does not become more human; rather the human becomes more machine. The mother's body turns like cog and wheel, rendered senseless by anesthesia—but not invincible by it. Her body may be torn asunder, sewn back together, and so on with the same gut-wrenching collision of flesh and tech we see in *Whitechapel Gods*. The worst of our fears exist at these borderlands, at the site of intersection, where we see, but cannot tell the difference.

## *Technology Becomes Us*

Descartes and La Mettrie, Vaucanson and Jaquet-Droz, William Blakey and William Smellie—each sought to reorder the complex machinery of being human. That they were not entirely successful might be gathered from our own strange and wondrous bodies, driven by purposes we scarcely understand and rarely control, unavoidably aging and in a constant state of incomprehensible change. We march toward death, though all of medical knowledge chases after the elusive mathematics of life, holding out some hope through gene sequencing and DNA that we'll arrive at last to the sacred text Dr. Culpeper thought he'd already mastered: the cosmic map of *us*. We might chuckle at his model, but we so often see what we expect to see—and, as Descartes himself would remind us—none of it may be reality, if any such thing exists. He spoke of the body as hydraulics and pulleys; later, La Mettrie would speak of the mind in much the same way. Still later, Thomas Willis would give us our first idea of electrical nerves, and since the advent of computers, we find ourselves talking about brain power as a matter of circuits. But none of us is correct. "A machine, any machine," says Elting Morison, "tends to establish its own conditions, to create its own environment and draw men to it." That is, we bend to the new order because machines can be designed to do only a part of what a whole man can do—and so "wear[s] down those parts of a man that are not included in the design."[56] We built human machines to demonstrate an order we expected to be there, in the way that Kepler built machines to show that the universe itself corresponded to a text we could read, a riddle we could decipher. But that always anticipates the order to be found. And because the technology we invent irrevocably changes the way we see the world (and even which and what parts of worlds we see), the more ordered the machine becomes, the more we expect all things to be ordered *just like machines*. "Man is, not only because he thinks," Elting explains, "but because he feels."[57] Descartes has it only partly right; we may be, of all creatures, the most intellectual and certainly the one most taken with our own cognitive abilities, but whether you house the immaterial in Descartes's disembodied mind or La Mettrie's organized parts, we remain a composite being, a "creature of rapture and despair,"[58] Mama Engine and Grandfather Clock in a palpitating, organic, engine of becoming.

Today, this strikes at the heart of the android project, undergirds the "emotion" machines and empathy robots already in design, if not in production. The fondest hopes of AI and the darkest dread of science fiction dystopia—the clockwork boys and the mother machines—linger in doubt over the same, singular issue: to build the human machine, we must first understand the human. To understand the human, we built machines. The inventor, in fiction and in fact, operates first as a unit of disruption, a wrench in the works who refuses the status quo (and even sensible defeat).

The century that followed Newton and Descartes built upon their scaffolds. Chaos had not been overcome, but rather managed, subsumed, made part of the works and systems. Out of the chaos, we imposed order, a constructed reality—even a *Matrix*. And from those battleworks, the next generation sought to throw the doors off of Nature itself, and to the darkness, bring an increasingly *scientific* light. They would do so in contest, too, fighting for the right to be first, to be best, to be remembered beyond the grave.

# PART TWO

"I try in vain to be persuaded that the pole is the seat of frost and desolation; it ever presents itself to my imagination as the region of beauty and delight. There, Margaret, the sun is forever visible, its broad disk just skirting the horizon and diffusing a perpetual splendour. [. . .] What may not be expected in a country of eternal light? I may there discover the wondrous power which attracts the needle and may regulate a thousand celestial observations that require only this voyage to render their seeming eccentricities consistent forever. I shall satiate my ardent curiosity with the sight of a part of the world never before visited, and may tread a land never before imprinted by the foot of man."

—Letter I, Mary Shelley's *Frankenstein*

# THREE

## Catching Lightning in a Bottle

W hen George Shattuck Morison stood before an eager crowd of would-be engineers to announce the new age, he already lived in the future of power. Everything the nineteenth century had been, and all that the twentieth century aspired to be, depended upon that power, and upon the dark seams of coal that fueled it. The New Epoch was made of *old death*, the fossilized remains of onetime creatures, the vegetable and animal life of eons offering an industrial birth out of coal tar and soot ash. But this promise of resurrected life also sits at the heart of a singular story—one that still haunts us, suggesting with eerie composure that life may spring from dead tissue, that we may command life's magical force. Mary Shelley, the daughter of William Godwin and Mary Wollstonecraft, and the lover and later wife of poet Percy Shelley, wrote *Frankenstein* in the dread winter weather of 1814.

The novel, like Frankenstein's creature, is a composite body, built from hopes and dreams of a previous century when electricity itself was still new, a mythic fire controlled by the gods. The novel, also called *The Modern Prometheus*, begins by shaking fists at that Creator-God, borrowing a quote from Satan in *Paradise Lost*: "Did I request thee, Maker, to mould me?" Did I ask, in other words, to be born of this clay and heat and fire? Victor Frankenstein never gives the game away, but the novel's greatest foreshadowing comes in the form of a lightning strike. Lightning, the first of our dreams of power, destroys indiscriminately, behaves unpredictably. The tongue of flame splits the darkness, and men would harness it—literally—by catching lightning in a bottle. Even so, power didn't consist entirely in *light*, but in *enlightenment*, in turning the lens away from the heavens and toward the earth's own distant corners.

Shelley's novel does not begin with the scientist's laboratory but instead with the explorer's ship, with Walton writing letters home about a country where there is *no darkness at all*. The "new philosophy" of science had spawned a new kind of "philosopher." The Enlightenment, a European intellectual movement that spanned the seventeenth and eighteenth centuries, relied on reason and rationality and method. It may have grown from thinkers like Newton and Descartes, but Enlightenment science privileged practical ends, observable methods, explainable principles, and a universe free from superstition—and not infrequently, from God. The gentleman of means experimenting at leisure slowly gave place to the engineer and businessman of which James Watt and Matthew Boulton, creators of the first effective steam engine, make perfect examples, and also to the botanist and explorer searching far-flung frontiers. "We cannot conceive of any end or limit to the world," said Francis Bacon, "but always as of necessity [. . .] there is something beyond." We can scarcely find a place on earth today unlit by neon and streetlamp or at least the pink haze of far-off cities—and we can barely imagine a world with "undiscovered" countries. The eighteenth century had both, a vast and limitless unknown to which intrepid inventors and explorers would bring their separate kinds of "light."

While it might be common to think of steam power coming *before* electrical power in a linear progression, history tells a different story. In every way possible, men sought to make and control power. Hydraulic power, steam power, and the fiery possibility of electrical power enticed equally—and early. Long before steam engines ran upon Victorian rails, electricity fascinated Benjamin Franklin in a young America. And long before Ben Franklin envisioned his first storm kite this same mysterious power drove a controversy over life force between Allessandro Volta and Luigi Galvani, resulting in the discovery of the battery on the one hand, and a series of grisly experiments on the other. The century's explorers sought a different kind of "enlightenment," new worlds to discover that required wooden ships and steel determination, and which frequently led to the starvation and violent death of the explorers themselves. *Frankenstein*, in all its sprawling, stitched, and grim-lipped splendor, encompasses both the age of electricity and the age of "dark continents"—the power of life and the power of death. But first comes the spark.

In January of 1746, Pieter van Musschenbroek reported the following to colleagues at the Académie des Sciences in Paris: "I would like to tell you about a new and terrible experiment, which I advise you never to try yourself, nor would I, who have experienced it and survived by the grace of God, do it again for all the kingdom of France."[1] Musschenbroek had been experimenting with a device for "storing" electrical "fluid," but he got much more than he bargained for: "My right hand was struck with such force that my whole body quivered just like someone hit by lightning [. . .] the arm and the body are affected so terribly I can't describe it. I thought I was done for."[2] As experiments go, it was, in his own words "explosive." But the shaken professor concludes in despair, "I've reached the point where I understand nothing and can explain nothing." He'd had the nearest experience one can reasonably get (and live) with electricity, but he had no idea how it worked, or why *this* experiment, of hundreds he'd performed, had worked so violently. Sixty years later, Humphry Davy would demonstrate this same strange fire to the Royal Institution—but in perfect command of the laws and convinced of the possibilities for lighting the future. Between these events, wonder-workers, charlatans, "electricians," philosophers, and countless experimenters whom history has forgotten played a dangerous game with the most powerful force yet known to man.

## A Troubled History

It's no wonder the Ancient Greeks deified lightning strikes as weapons of Zeus, an armament that would split trees, rend earth, and destroy men and beasts with falling fire. The ancients did not, however, call the lightning *electricity*. Instead, the word comes from polished amber (ἤλεκτρον or *elektron*). Formed of fossilized tree resin, amber fascinated because of its color and texture, but also because it could be *charged*. Though a "stone," amber feels strangely warm to the touch having gone through a process of molecular polymerization. A Greek philosopher circa 500 B.C.E., Thales of Miletus, discovered that rubbing amber with a cloth caused it to discharge a spark, and in 1671, Newton's rival, Leibniz, discovered that sparks were associated with electrical phenomenon. Even so, for decades to come, no one would consider the humble static shock and the terrific bursts of thunderstorms to be one and the same. When Newton finally took the reins of the Royal Society in 1703, the first experiments had to do with air pumps, vacuums, and static generation, all as man-made (rather than God-made) wonders. But it would be the static generation machine that captured public imagination; the modern world owes a great debt to a draper's son turned lab assistant and chief experimenter of the Royal Society: Francis Hauksbee.

Hauksbee began his life in the trades, in Colchester, England. He'd been apprenticed to his elder brother, but spent his time researching the experiments of the new philosophers. By 1699, Hauksbee had left off the fabric business and, like William Blakey and his clockwork trusses, turned his skills toward the manufacture of medical devices.[3] He advertised "cupping" apparatus to the public—bell-shaped glass devices that could, with the aid of suction and depression, create a vacuum against the skin. In addition, Hauksbee experimented with engines for creating pressurized air, pumps, and barometers—all for practicing physicians. We have very little evidence of their medical efficacy, but we can speculate on one thing: through his advertisements, Hauksbee gained the attention not only of the physician members of the Royal Society, but of Newton himself. On the fifteenth of December, 1703, a then-elder Newton invited Hauksbee to revive the society's demonstrations—and a few years later he invented the most curious (and arguably most famous) of his "machines."

A glass sphere with the air vacuumed out, a wooden support, a belt, a crank, and a wheel [Fig. 6]: standing before the waiting Society members, Hauksbee asked for the candles to be snuffed. He rapidly turned the crank, the spinning sphere clicking along in oscillating cadence. Then, in the hushed dark, he placed both hands upon the glass. Where only darkness had been, a dancing, ephemeral blue light emerged. It danced about his fingertips like a living creature, collecting, dispersing, a ghost light. Hauksbee's machine generated static electricity on command, the very first mechanical generator ever invented. "The Learned World," Hauksbee wrote in 1709, "is now almost generally convinced that instead of amusing themselves with Vain Hypotheses [. . .]."[4] His "New Experiments" would not only demonstrate the reality behind Newton's theories on light, they would explore the "Laws of Electrical Attractions," that is, *electricity* and *magnetism*. More importantly by far, however, they invited others to do the same. Hauksbee published diagrams of his experiments, making them both accessible and replicable, and soon shocks were being generated for excited crowds at dinner parties. Self-proclaimed "electricians" would raise feathers with static-charged glass rods and even set fire to glasses of cognac with a spark from their charged fingertips [Fig. 7]. But Hauksbee didn't take part in the further demonstrations, no matter how outlandish; the Royal Society lost interest in what they considered parlor tricks for street magicians. It wouldn't be until the 1730s and '40s that their attention, with that of a wider world, would be reignited for these curious blue sparks.

Stephen Gray had been a silk dyer before an accident rendered him disabled. He became a pensioner at Charterhouse (for orphans and elderly), but Gray had an active mind and an interest in electricity. Now, with time on his hands and plenty of useful orphans about, he began a series of unusual, and not entirely *safe*, experiments. Gray suspended one of the youngsters by silk ropes in the great hall and charged his prone body using Hauksbee's machine. Later called the "flying boy" and reintroduced to those parlors and back rooms as excellent and "innocent" fun, Gray's swing-like contraption proved the next leap into electrical understanding. Electricity *traveled*. It could move from the machine to a rod, the rod to the boy, and from the boy's fingers to a plate of feathers or gold foil that danced and sparked beneath him. Why the silk ropes? He must be suspended above the ground for it to work—and silk did not allow the electricity to move through it. Soon Gray was conducting other experiments with the aid of his friend Jean Desaguliers. A cork, for instance, could be electrified by means of hemp thread suspended on silk *though nine hundred feet away*, the same principle we use today for sending power through high-tension lines, the pulse skirted along wires through insulators to keep it from escaping. The world could be separated into *conductors* and *insulators*; electricity stopped dead in silk and resin, but it was conducted easily by metals, fluids, and—disturbingly— human bodies. Gray's ideas whipped up national and international attention. Itinerant lecturers toured capitals and provinces with portable Hauksbee machines, offering demonstrations in public squares and aristocratic salons, with the human body at the center of experiment. Their activity, says historian Paola Bertucci, made electricity "one of the most discussed topics of polite conversations." In 1745, readers of the *Gentleman's Magazine* would hear of "wonderful discoveries, so surprising as to awaken the indolent curiosity of the public," even for "ladies and people of quality, who never regard natural philosophy but when it works miracles."[5] The Newtonian maxim of preserving God's secrets for the chosen had been overrun by an increasingly literate public of pleasure-seekers. No longer God's thunder-bolts, electricity was, as the *Gentleman's Magazine* explained, "new fire which a man produced from himself, and which did not descend from heaven."[6] On the wings of rational intellectualism, electricity traveled fast, but it still could not be *contained*. How, the wonder-workers asked, could they make the spark brighter and more long lasting? Could it be contained and saved?

The problem for "electricians" and natural philosophers studying electrical effect had to do with storage. Stephen Gray could do wonders with the "flying boy," but the charge wore off when the sphere ceased rotating, dissipating and disappearing as it went to ground. Those who wanted to study it struggled to find useful means, especially those, like Professor Musschenbroek, who wanted to demonstrate the principles for his students at Leiden University. Because experimenters assumed that electricity was fluid (sensible, since it "flowed" from person to person, or person to object), they theorized it might be captured in a glass jar. Musschenbroek and his assistants assumed that placing a wire into a jar partly filled with water (a conductor) and then placing the same on a resin insulator, they could use the Hauksbee machine to capture electrical charge inside the glass. The theory was sound—but it didn't work. No matter what they tried, they could not generate a spark. Not, at least, until Musschenbroek, tired from lectures and probably in need of a late dinner and a good sleep, accidentally performed an experiment *the wrong way*. Musschenbroek charged his jar while holding it in his hand and without using the insulator. After cranking the Hauksbee machine, Musschenbroek touched the top of the jar and its copper wire—and had the shock of his life. In the letter, he claims that he would not try the experiment again, but his caution gave way to curiosity. News of the device, its curious claims, and its horrific powers spread from Holland to England, then to distant Japan and also to the American colonies. The Leiden (or Leyden) jar worked like a battery and an early condenser. And despite or, as historian James Delbourgo suggests, *because* of Musschenbroek's cryptic warning, the Leiden jar soon "became an irresistible object of both philosophical curiosity and spectacular corporeal experience."[7] And still, no one knew what it was, or why the human body itself seemed so great a part of the electrical force.

Electrical phenomena "as simple as attractions and repulsions between charged bodies" were choreographed "to keep audiences from boredom," says Bertucci.[8] Spectators could "feel" the effects and instrument makers after Hauksbee exploited what we now know as "electrostatic induction" to make paper puppets dance. The Bakken Museum has one such device; tiny figures, light as air, would skim and pivot on the static charges of metal discs [Fig. 8].[9] To hold that history in your hand, as I have done, is to take part in that fascinating display of theatricality. Think how wide the eyes, how hushed the voices. Think how astonished we would be, ourselves, to see the magic of sparks for the first time. It would be Franklin himself who eventually unlocked the secret of the Leiden jar's force; he took to making his own experiments and—possibly because of his own background in business—developed an idea of electric *economy*. There must be positive and negative charge, he reasoned, and because nature seeks always to balance her books, one rushes to cancel the other. In 1747, shortly after the publication of Musschenbroek's findings, he established the principle we take for granted today: the Hauksbee static machine *did* charge the jar (itself a non-conductor), but with a negative charge that dissipated. However, the human hand placed on the jar, and the human foot touching earth, sent a positive charge to the glass surface, where it steadily built up. The negative charges built too, on the other side, but neither could travel through the glass, and thus one would still feel nothing. It would take the other hand touching the uninsulated lid to complete the circuit—and the sudden rush of charge from negative to positive resulted in explosive shocks.[10] To his critics, Franklin enthusiastically offered "proof": "If anyone should doubt whether the electrical matter passes through the substance of bodies, or only over the surfaces, a shock from an electrified glass jar taken through his own body, will probably convince him."[11]

Bodies attract. Having made a fair number of unfortunate mistakes around electric devices, I know how readily flesh and blood receives and conducts. There is nothing amorous, nothing enticing, nothing rapturous about the shock of balancing electrons. But the inclusion of the body in eighteenth-century electrical demonstrations aroused more than curiosity. By 1749, the electrical showmen had secured audiences also in Philadelphia, and Benjamin Franklin jested of electric parties where "a turkey is to be killed for dinner by the electric shock, and roasted by the electric jack, before a fire kindled by the electrified bottle."[12] Women became "essential protagonists of electrical soirees," and spectacles played with sexual difference.[13] One famous trial coaxed young men into kissing electrified young ladies. A poem published in 1754 described the experience as one that "pained me to the quick" and almost "broke" teeth in the shock. But hopefuls line up anyway, somehow more attracted by the fact they could not reach the object of desire—some more than once. In the dark of aristocratic salons, in private drawing rooms of the merchant class, in the great halls and public parks of the masses, electrical performers staged a host of bizarre sensory experiments: electric fire could be seen in the dark, it could be heard in the crackle of discharge, it could be smelled in the sulfur tinge (and occasional burnt hair of participants), and best and worst of all, it could shatter through the body itself, shaking limbs and rattling teeth. Seeing, but also *feeling*, was believing: we had discovered *the body electric,*\* our own blood and bones somehow provided a channel for the mysterious electrical fluid. And for reasons that defy common sense, the dangerous force that so unsettled and damaged Musschenbroek came to be thought of as the latest in medical therapy. The spark of light might just be the spark of *life*.

---

\*   From the 1867 version of Walt Whitman's *Leaves of Grass*: "I sing the body electric/ The armies of those I love engirth me and I engirth them/They will not let me off till I go with them, respond to them/And discorrupt them, and charge them full with the charge of the soul."

## *Animated Beings*

*"As I stood at the door, on a sudden I beheld a stream of fire issue from an old and beautiful oak which stood about twenty yards from our house; and so soon as the dazzling light vanished, the oak had disappeared, and nothing remained but a blasted stump."*
—Mary Shelley, *Frankenstein*

How often have we looked to the heavens during a storm and wondered at the fractured bursts? The sizzle of white pulses like cracks in the roof of the world. No one could master that power, no one but God—and men, mere mortals, should not try. Hollywood has rendered the making of Frankenstein's creature a moment of electric hubris, a taming of heaven's fire to imbue a monster with life, but Shelley's story gives only the barest hints. When Walton, having nursed and befriended the dying Victor Frankenstein, asks for particulars, he is rebuked. "Are you mad, my friend? [. . .] Or whither does your senseless curiosity lead you? Would you also create for yourself and the world a demoniacal enemy? Peace, peace! Learn my miseries and do not seek to increase your own."[14] Even so, Frankenstein begins with the blasted oak reduced to ribbons, a moment that affected him as a light dawning in darkness. Here was power. Here was something he might tame, even life itself. The Royal Society had turned its back upon electricity as a pastime without utility; they had missed the opportunity to see with fresh eyes what might be possible, and (particularly under Newton) preferred the mysteries of God to the systematic testing of the elements themselves. That task would be left to a man at odds with the very system on which the Royal Society was built. Possessed of "innate common sense," honest empiricism, secular practical utility, and civic-minded benevolence"[15] (or so goes the common mythology), Benjamin Franklin represents a revolutionary idea: science is for everyone—not the elite, most particularly not the elite of England.

A colonial, and soon to be a democratically minded one, Franklin's philosophical, political, and rational position refused superstition of any kind. He is widely cheered for flying a kite in a thunderstorm (something he probably didn't do), but the most fascinating experiment he devised had nothing to do with kites and copper keys . . . Neither did it take place in America. In 1752, Frenchmen George-Louis Leclerc and Thomas-François Dalibard followed Franklin's theoretical designs and erected a forty-foot metal pole held in place by wooden staves.[16] The pole rested inside an empty wine bottle. Franklin had already invented the lightning rod, and recommended that they be fitted to churches and other tall buildings so "the electrical fire would [. . .] be drawn out of a cloud silently, before it could come near enough to strike."[17] His theory for the Leclerc and Dalibard experiment was that the rod would "catch" the lightning and store it in the bottle, just like a Leiden jar. On May 23, at 12:20 P.M., lightning hit the tip of the pole with incredible force, but the bottle remained intact. An assistant ran forward, and as he neared, a spark leapt from the bottle to his hand, burning it. The force may have been greater, but Leclerc and Dalibard confirmed Franklin's rationalist theory—lightning and man-made electricity were the same. More importantly, man could "catch" lightning and, as with the lightning rods, render its powerful force (relatively) harmless and contained.[18]

"The untaught peasant," complains Victor Frankenstein, "beheld the elements around him and was acquainted with their practical uses. The most learned philosopher knew little more."[19] But with the advent of new science, even the elements could be controlled. Men might "penetrate into the recesses of nature and show how she works in her hiding-places. [. . .] they can command the thunders of heaven, mimic the earthquake, and even mock the invisible world with its own shadows."[20] By the 1750s, Immanuel Kant hailed Franklin, not Frankenstein, a "modern Prometheus," stealing heaven's fire and giving it freely to humankind. The tale makes up part of the American story; a mere colonial, humble and democratically minded, tames the natural world while British elites look on in superstitious wonderment. He stands as the best representative of the Enlightenment natural philosopher in an age where "empirical demonstrations of cause and effect" showed mastery of nature, and as with the clockworks of the foregoing century, an understood order.[21] But the public's chief interest in electricity, and one that occupied Franklin, too, in the years leading up to the American Revolution, had everything to do with bodies. In the 1740s, German doctor Johann Krüger and his pupil Christian Kratzenstein noticed that electricity caused involuntary muscle movement.[22] Incredible claims of miraculous healing soon followed in magazines and even respected journals, everything from gout, baldness, paralysis, nervous conditions, circulatory distress, and more supposedly "cured" by miracle devices. The stories were rarely substantiated and most patently false; the authors of *Epitome of Electricity and Galvanism* complain of this charlatanism and "vulgar amazement," warning that "when wonder and credulity are coupled with terror and surprise, we must look for a strange and misshapen progeny."[23] Small sparks didn't just light the room; they made the surrounding darkness all the blacker. What had been achieved in the eighteenth century was, as Frankenstein describes it, only "the partially unveiled face of Nature." Like a well-oiled machine, "one by one the various keys were touched which formed the mechanism" of Victor's being; "chord after chord was sounded, and soon my mind was filled with one thought, one conception, one purpose."[24] Though operating with scant information, and most of it wrong, Shelley's mad scientist awakens to the new possibilities: "I will pioneer a new way, explore unknown powers, and unfold to the world the deepest mysteries of creation." At the end of the eighteenth century, that mystery would be called "animal electricity" and would lead to a debate as vehement and as influential as the one between Newton and Leibniz over the providence of God. *Modern Prometheus*, indeed.

Luigi Galvani, Italian physician, biologist, and philosopher, esteemed professor at the Catholic University of Bologna, spent his evenings closed up in tight quarters with old books . . . and skinned frogs. Franklin suggested flying kites, but Galvani's lightning experiment attached wires from frog's legs to iron rods during a storm, just to watch them kick. It began with an accident; he and his assistant prepared frogs for a static experiment—and when the student touched the metal blade to the frog's sciatic nerve, the legs jumped as though in life (despite the fact that the frog's head had been removed). It's an experiment that may be repeated in any lab in the country that might have the tools of high school science. We *know* the cause so intuitively that it's hard to step back from it, as hard as imagining the world as flat and the sun spinning around it. But consider the dark night, the dim space lit by guttering candle; a sulfurous fume and the bitter metallic of electric energy dancing on the tongue. A creature entirely disemboweled, piecemeal, scarcely resembling the thing it had been in life, suddenly animates. Its toes splay and grasp, the muscle firing to jump, to flee. Lightning and static may be proved much the same thing, but what was this strange power that emanated from the body itself? It called to mind another strange phenomenon, the "sting" of the torpedo fish and the eel, described by Henry Cavendish some years before.

Cavendish became convinced that the fish fired an electric charge to kill its prey, and the reports of all who had experienced the sensation confirmed it to be the same in nature as electric shock. At the same time, the fish created no spark . . . and neither did the amphibious limbs twitching on Galvani's work table. After multiple experiments, he discovered that insulators and weak conductors (set apart by Stephen Gray) didn't cause a reaction, but use of metals *did*—and it reminded him of something: "when the phenomenon of contraction occurred, the flow of very tenuous nervous fluid from nerves to muscles resembled the electrical flow discharged from a Leyden jar." He even put himself in the "circuit," just as he'd learned from Franklin's experiments of positive and negative charge.[25] It worked—just as the torpedo fish could shock without recourse to a Hauksbee machine, the frog's muscle and tissue seemed to hold a magical property deep within. His lab became home to more jars and filaments, speared tadpoles, electrified limbs, nervous systems with the organs stripped away, and rows of tiny bodies dangling from copper wire like macabre Christmas decor. He called the magic substance "animal electricity," or—in honor of Galvani—*galvanism*, and it became the most contested concepts of the Enlightenment. The old world with its ideas, its mysticism, its giddy promises was about to collide with the new science, and nothing (least of all Galvani himself) would ever be the same.

Victor Frankenstein discovers galvanism during a lightning storm; he claims, from this moment, a sudden awareness that his old masters, the dusty tomes of the alchemists, were wretchedly flawed. At the same time, galvanism didn't offer better answers, just new mysteries. "All that had so long engaged my attention suddenly grew despicable," he recalls to Walton, and "set down natural history and all its progeny as a deformed and abortive creation, and entertained the greatest disdain for a would-be science which could never even step within the threshold of real knowledge."[26] It recalls the words of the *Epitome of Electricity and Galvanism*, published a few years before Shelley's novel, almost verbatim: the dangers of wonder and ignorance haunt Frankenstein for the same reasons that electricity tempted the scientist and also the credulous, horror-struck crowds. Math might have lead Kepler and Newton to great heights, but it would never do for the carnival. Who would shout the glory of calculus from a circus tent? Electricity, however, worked by principles still unknown even to the trained mind. It danced. It sparked. It threatened life, but it also—seemingly—*gave* life, from the twitch of paralyzed muscle to the kick of a dismembered frog. Animal electricity was life itself. Galvani did not ask whence it came; for him, God had put fire into the very stuff of bodies, imbued it with animation and with life. Man might create a false kind of shadow, just as he might build automatons that replicated but could not replace the *real*. But he had rivals too.

At the start, Galvani's 1792 work, *De viribus electricitatis in mortu musculari* (The Effects of Electricity on Dead Tissue), was met with intense excitement. No longer was electric pulse only to be found in eels and torpedo fish; instead, it was a part of all beings, a vital fluid that accounted for all the activities of life—a latter-day alchemy, a magic bullet. Here was life, soul, and the animating principle in one. Other well-known electricians took up Galvani's cause, including Tiberius Cavallo and the physiologist Richard Fowler. Fowler trained in Edinburgh and had his work published in London. The Dittrick Museum and Allen Library in Cleveland retain a copy from 1793; it's been rebound, but its size made it perfect for carrying in Enlightenment coat pockets. Easy to read and full of exciting firsthand accounts of Fowler's own frequently strange experiments, *Experiments and Observations* offered a different vision from Galvani's Latin treatise. Fowler's book opens (like this one) with an accidental discovery and an unfolding tale:

> *Some accidental appearances, certainly electrical, excited, by their novelty, the attention of the Professor of Anatomy at Bologna, to the investigation of the possible, but unknown, dependencies [. . .] of animals upon electricity; and the astonishing effects of that influence upon the human body.*[27]

Fowler goes on to describe his own attempts at following Galvani's method, using zinc, nickel, gold, copper, and silver to excite the contractions in the muscles and nerves. The book ends with experiments for the reading public to try, if they have the stomach for it: "as evacuating the blood from a living animal is rather a severe operation," Fowler explains, it might be more useful to "crush the brains" before proceeding. "I injected [. . .] thirteen drops of opium into each of the hearts of the frogs [and] the hearts became white and ceased contracting"—forty-eight hours later, Fowler attempts to excite contractions in the mangled, bloodless amphibians, but the bloodless limbs produced only the slightest action.[28] He continued these daily sequences until the frogs became putrid and foul smelling, then he begins again with his next victims—among them rabbits and dogs. Fowler does not, in the course of his experiments, move higher up the great chain of being, but Galvani, it should be remembered, was also an anatomist. "To examine the causes of life," Victor Frankenstein explains, "we must first have recourse to death [and] observe the natural decay and corruption of the human body."[29] Galvani meant to show the "how" of electricity; he did not intend to question the divine *"why."* He was a man of Newton's ilk. Fowler used method and experiment to prove Galvani's claims, but charlatans and public demonstrators capitalized on the ambiguity. In 1798, a "rational mystic [. . .] being the true and lawful Heir of PROMETHEUS" had written to explain electricity is "the vehicle of thought, and peradventure of the human Soul itself."[30] Again, the references to Greek gods and divine fire, but these enthusiastic responses had more of superstition to them and less of science. They would have been decried by Franklin (who died in 1790), and they would not be countenanced a hundred miles away in Pavia, a city swept up in the European Enlightenment. But the idea that electric power contained the secret of life itself caught fire anyway—an antidote to chaos in its very luminosity.

## The Modern Prometheus

Galvani did not ask where the spark of life originated, but Victor Franken-stein does. On the cold decks of an icebound ship, its sides lashed by gales that caused timbers to rock and groan in plaintive, splintering squeals, the young Walton prepares to hear the strangest of all stories. But despite how Hollywood has chosen to portray the "mad doctor," Shelley's original narrative gives us, instead, an enlightenment natural philosopher. "In my education," Victor explains, "my father had taken the greatest precautions that my mind should be impressed with no supernatural horrors."[31] He never once trembled at a tale of superstition or feared ghosts and spirits. Walton sets a course to the north where the sun never sets, but upon Victor's fancy, "Darkness had no effect" and a churchyard was "merely the receptacle of bodies deprived of life." He spends his days and nights in such charnel houses and among the collected bones of the dead and gone, and he thinks what a terrible waste of nature's most perfect creature: "I saw how the fine form of man was degraded [. . .] I beheld the corruption of death succeed to the blooming cheek of life; I saw how the worm inherited the wonders of the eye and brain."[32] Victor may not fear the darkness of death; he intends, instead, to overthrow it. Darkness is anathema. He would bring the light of science (he tells us), the patient, methodical, plodding light of painful, dull, repetitive work. His hours of study and toil in the boneyard and anatomy theater are ignored by film and screen, but for the Enlightenment scientist, the method is the point. Work in tiny increments, little by little, and in conversation with other thinkers—that is how light dispels the darkness. Frankenstein worked alone, but real-life scientists rarely do. Shortly after Galvani published on animal electricity as the beginning of all life, commentaries on the work are delivered into the hands of one Alessandro Volta. Unlike Galvani in almost every respect, Volta reads the discoveries and, unable to stomach the grand claims, sets about testing them by his own methods.

Alessandro was to Galvani as Leibniz had been to Newton. He appreciated the good things in life, dressed well, and was reasonably attractive and popular with the ladies. He liked high living, drank too much, and kept current on the latest operas—and, like the other "genius" temperaments described by Elting Morison (or the occasional steampunk supervillain), Volta thought the world of his own unique and powerful mind. To better himself, he applied to learned men with a combination of fervor and charm, and so built a social network that eventually earned him a place at the University of Pavia. He could both flatter and please, and, explains literary scholar Roseanne Montillo, his "knack for showmanship" gained him both peers and patrons.[33] In Newton's day, Leibniz could be outfoxed by staunch stratagem and sour-faced loyalists to order and restraint. At the century's end, the tables had turned. Franklin and the electrical experimenters had all been showmen, to a degree, and electricity offered the first best example of a science that lent itself readily to the public. But showmanship also launched balloons.

The dream of flight, says Richard Holmes, enticed humankind since the "myth of Icarus" (who fell to his death when his wax wings melted in the sun).[34] And flight, like lightning, was a thing of the gods. It should surprise no one that Benjamin Franklin, at the end of his life, spent his time reporting aerial experiments to the Royal Society. Two Frenchmen, Joseph-Michel Montgolfier and his brother Jacques-Étienne, launched a hot-air balloon—and others soon followed. In 1785, the imposing figure of a living Juno—a woman of substance in low-cut silks—boarded a striped and stately balloon basket at Hyde Park in London. The aeronautical master of the occasion, the Italian Vincenzo Lunardi, gave the day its flourish. Dashing, gallant, fearless, and proud, Lunardi had become the toast of England for his expeditions as the self-proclaimed "First Aerial Traveler in Britain." His accomplishments included being one of the finest marketers of his day, and many were the snuff boxes (and garters) graced by his name or form, or the figure of a balloon. Never mistaking an opportunity for a "first," Lunardi planned to launch the first Aerial Female—with her sashes and petticoats, good humor, and not a little courage. It sold tickets. It garnered support. But like the electrical experiment, flight was wretchedly dangerous. French aeronauts Jean-François Pilâtre de Rozier (another dashing national hero) and Pierre Romain fell to their deaths in an attempt to cross the Channel from Boulogne to Dover. The upper hydrogen balloon caught fire and they plummeted, Pilâtre leaping from the basket amid smoke plumes. The bodies of both were horribly mangled by the organ-rupturing impact. Despite all the carnage, the expense, the sheer danger of unpredictable winds and unknown landing sites, when the seventy-seven-year-old Franklin was asked about the use of a hot-air balloons, he scoffed: "what's the use of a newborn baby?"[35] Science and spectacle and wonder mingled, "misshapen progeny" or not. Volta knew the game better than anyone; did Galvani think he had found a secret? Volta would bring the fight to him in an arms race of experiments the likes of which Victor Frankenstein had never seen.

At first, Volta was inclined to admit animal electricity as a possibility. He'd done his own experiments and measured weak electricity from frogs, but Volta could not shake the idea that the frogs' bodies were passive, rather than active, parts of the electrical process. We can peer in on Galvani's method even today; his experiments, macabre as they were, have been well documented and well preserved [Fig. 9]. He exposed the nerves of frogs, including the sciatic nerve. He then touched the nerves with metallic rods or pins. The original experiments included the use of a Hauksbee machine, but Galvani repeated it (as did Fowler) with all sorts of varying implements of zinc and silver, and also copper. The nerves, he said, were conductors—the muscle the "Leiden jar," storing it up to be released with the circuit was closed with two different types of metal. But even Fowler noted that not all metals worked the same. And a new idea nibbled at Volta's understanding. Perhaps the *entire frog* was a conductor, and the electric current had been generated from the metals themselves. He made a long wire of two types of metal and pressed the ends to a nerve. The muscle contracted, though not part of the "circuit"; it must be the dual metals, acting together as a "biometalic arc."[36] Now it was Galvani's turn. He and his assistants tried new tests with different frog preparations. This time, they left the metals out of it and merely touched one part of the frog to another part, nerves to nerves.[37] A strange sensation occurred, and a kind of acidity followed. He had "tasted" electricity—and he followed the first experiment with one more nerve-shattering (and worthy of Newton's self-experimentation): *he put the arc contacts into his eye.* The result was unexpected. In the darkness—there was light.[38] Not a burst, not quite, but a pinprick of light that registered to Volta how the electricity worked.

The result of the metal-free experiment sent Volta back to his lab, but not in defeat. An elastic mind and complete faith in his abilities anchored him while he searched for a solution. The answer, he decided, had to do with "humid conductors"; Galvani had touched one slick, wet, skinless part of a frog to yet another slick, wet, part. Somehow, the electric fluid communicated in that environment. But there was no way to prove it, not from that quarter. If Galvani determined to take the metal out of the experiment, Volta would remove the body itself. He'd already done a test of the dual metal wires by pressing them to his tongue. Now he set about creating a false frog, a construction entirely of those metals and weak acid (which the fluid of the mouth and eye represented) in a "pile." Layer by layer, he built the stack of metal and acid, attached the arc wires, and presented the first electricity that was neither mechanically generated nor emanating from some "vital fluid" inside biological flesh. But the strange, messy, sulfurous, and bubbling device did something else too. It produced continuous electric current. Leiden jars must be charged; they stored only for a short time and produced electricity in bursts. The pile, by contrast, offered something entirely new. Volta had invented the *battery*, and nine years later, a young Humphry Davy would take the stage to astonish the Royal Society with the experimental science they once turned their back upon: steaming stacks of Voltaic piles lined a basement room, and above stairs, he delivered the first arc lamp demonstration, bright, white, blinding and *constant* light. And still the question remained: what was this substance, and what could it be used for?

"From the midst of this darkness," says Victor Frankenstein, "a sudden light broke in upon me—a light so brilliant and wondrous, yet so simple, that while I became dizzy with the immensity of the prospect which it illustrated, I was surprised that among so many men of genius who had directed their inquiries towards the same science, that I alone should be reserved to discover so astonishing a secret."[39] The light, for Frankenstein, was *life*—the ability to reanimate dead tissue, to steal the beating heart of man from the grim clutches of death. This fascination, morbid though it may be, has been with humankind almost from the first moment we looked upon our wondrous bodies and realized they would not last. The morbid fascination remains with us still, shadowing our discoveries in science and medicine, dogging our steps. Volta won the scientific argument and vanquished Galvani, who died in Bologna a broken man with his faith shaken and his reputation in ruins. But Galvani's work returns in surprising ways; his nephew Giovanni Aldini sought to resurrect his brilliant, not by electrifying frogs, but by *reanimating dead humans*. The perfect solution to the human machine would be the ability to reengage it, to rebuild it, to achieve what only fiction had yet dreamed of.

On January 17, 1803, Aldini advertised the event of the age. The body of murderer Thomas Forster, hanged at Newgate, was brought to the public arena in London. A gruesome report follows—its detail tantalizing in its similarity to Frankenstein's. Shelley writes: "I saw the dull yellow eye of the creature open; it breathed hard, and a convulsive motion agitated its limbs [. . .] His jaws opened, and he muttered some inarticulate sounds, while a grin wrinkled his cheeks." In the report on Aldini's first attempts, "the jaw began to quiver, the adjoining muscles were horribly contorted, and the left eye actually opened [. . .] the fists clenched and beat violently the table on which the body lay, natural respiration artificially established."[40] Public outcry against these horrors resulted in them being banned and Aldini being chased back to Italy in 1805, but his experiments were taken up again by Johann Ritter in light of Humphry Davy's improved voltaic battery.[41] Ritter's work would be communicated back to the Royal Society with increasing uneasiness, and the final report ending with the chilling indictment: "it is impossible to conceive anything so disgusting and humiliating for the human understanding than their dreams."[42] For historian Richard Holmes, these dreams are impressed upon the very fabric of *Frankenstein*, not least because Ritter's led him to ruin—the loss of position, friends, family, and finally his sanity. To quote from *Hamlet*, "in that sleep of death, what dreams may come?" Mortality triumphed. Ritter died in 1810 still young, unaware that his ideas would be resurrected in a work we now consider both the beginnings of science fiction and the seedbed for steampunk tropes of mad science, technological hubris, and life's unfathomable forces. But its warnings didn't slow the march of progress into a disaster-blind future.

## *I Make the Light of Safety*

*"Life and death appeared to me ideal bounds, which I should first break through, and pour a torrent of light into our dark world."*
—Mary Shelley, *Frankenstein*

Around the same time Galvani and Volta battled in Italy, a whole other revolution was taking place in England. The German astronomer William

Herschel, inventor of the largest telescope the world had ever seen (at forty feet), discoverer of comets and planets, and—with Kepler—one who dreamed of mountains, and even cities, on the moon, delivered an astonishing paper about the cosmos. There was no God in the machine, he explained in 1791. There was *no machine* at all. For the first time, he calculated that the universe was far older than expected, and that it was subject to fluid changes over time. Nebulae and star clusters could be compared, explained Herschel, to plants in various stages of development, rather than mechanical contraptions.[43] The botanist Erasmus Darwin, grandfather of Charles Darwin, praised this new idea of an "evolving" cosmos. Darwin also honors Herschel's disturbing idea that the cosmos may collapse in the end, that "death and night and chaos mingle all."[44] In 1792, Herschel's friend Jérôme Lalande claimed, "I have searched through the heavens, and nowhere have I found a trace of God," while the French astronomer Pierre-Simon Laplace claimed the solar system required no special act of creation, no divine intervention.[45] An organic universe, constantly expanding and changing, would have devastated a mind like Newton's as the worst vision of chaos. But these new men of science, like Victor Frankenstein, saw the explosion of new ideas as the beckoning of uncharted territories. In 1802, Herschel developed the idea of deep space and deep time, the first inkling that looking at stellar light was looking into the past, seeing the distant ancestors of the cosmic universe, whose light had only just reached earth. He pushed his ideas outward, suggesting that there were other galaxies, even other Milky Ways.[46] The Romantic writer Lord Byron, friend of the Shelleys and present during the ghost-telling episode that spawned *Frankenstein*, visited Herschel's telescope as though visiting a strange and world-weary sage. The experience left him sobered: "the comparative insignificance of ourselves and *our world*, when placed in competition with the *mighty whole*," he confides, "led me to imagine that our pretensions to eternity might be *overrated*."[47] The sky for Herschel, explains Richard Holmes, was "full of ghosts," because "the light did travel after the body was gone."[48] Far from an ordered universe with an impenetrable but trustworthy God, the cosmos existed in flux with no one at the helm. The new breed of philosophers saw this great, mad world and invented new ways of pushing the boundaries. But even here, they chased the light. At the close of the eighteenth century, explorers and experimenters led risky, adventurous lives. They didn't want to stare into the void and despair—that wanted to hurl thunderbolts at it. And chief among them was Humphry Davy, already animated by Herschel's discoveries, and intent on illuminating the world.

Buoyant, attractive, and poetic in nature, Davy flirted with the Romantic ideal. Portraits preserve the sparkling luminosity of his eyes, a personality of vigor and verve that miraculously fit in his small frame (about five feet, five inches).[49] Mr. Davies Gilbert, who first employed him, described Davy's pleasure at being surrounded by scientific instruments as "the delight of the child introduced to a magazine of toys."[50] His conversation was excited and exciting, his personality charming, and his spirit adventurous to a fault. His experiments were often reckless (and smelly—the chemical odor a common complaint of his neighbors alongside the explosions of glass vessels unable to take the pressure of reaction).[51] Like Newton, he also experimented on his own body, even "forced exhalation" of air from his lungs using the pumps first designed by Hauksbee. After three tries, he passed out, but managed to mumble, "I do not think I shall die," to anxious assistants. And as early as 1801, while still a prancing young man (and man about town), his research into galvanism entitled him to a position with the Royal Institution of London. Best of all, the energy with which he applied his theories made his lectures popular not only among the learned but also the public. He is described in a 1868 tribute to "distinguished men" of science in Britain as possessing "a mind of a poetic cast," which supplied "a rich variety of metaphors and original illustrations, such as poets themselves regarded with admiration."[52]

The compliment is borne out in his earliest publications, a voice that sings with admiration of the sciences and of nature—but more particularly, of *light*: "Nature is a series of *visible images*: but these are constituted by light. Hence the worshipper of Nature is a worshipper of light."[53] Like the poets whose work spoke of cataracts sublime, Davy takes works from Kepler, Newton, and Descartes, as well as moderns like Herschel, Volta, and Galvani, and delivers them like a statesman's address: "We may consider the sun and the fixed stars (the suns of other worlds) as immense reservoirs of light, destined by the great Organizer to diffuse over the Universe organization and animation."[54] Mind and matter meld, the life force and its principles gathered together in what Davy calls the "Law of Animation"—a life spark that tied the mind itself to the "corpuscular motion," human consciousness and physiological process rarified to chemical constituents. "What had been the study and desire of the wisest men since the creation of the world was now within my grasp,"[55] exclaims Victor Frankenstein. Davy addressed new and larger crowds for his excited speeches, including one that a young Mary Shelley attended with her father. Davy, a mythic figure, was a showman of the first order who reveled in demonstrations of strange gases and chemicals; Davy proclaimed: "I was born to benefit the world by my great talents,"[56] a prodigy and a genius whose most sustaining work would really and truly light up the dark.

James Schafer and Kate Franklin's assessment of why "steampunk" science matters reminds us of the original brilliance of firsts, the bold open plain of technological discovery. Hauksbee's air pump had its ornaments, its dials and glass domes; his static generator whirred with gears and cranks. Franklin caught lightning in a bottle, Musschenbroek learned how to store it, and Volta how to generate it consistently. The greatest change in the history of innovation, says George Shattuck Morison, is the "manufacture of power," that is, in making, through ingenuity and accident, something really and truly *new*. Volta's battery had done this very thing, but the charge it created didn't amount to much: a mild shock. To convince the world, science needed a showman—and Humphry Davy was ready for the stage. On April 25, 1801, Davy delivered his first lecture for the Royal Institution on the subject of galvanism—a lecture unlike any that had been given before to the London elite. Holmes described him as "bouncing" onto the stage as a showman, launching his course without pausing for breath. Each experiment became a narrative, a story of hope and discovery. He must have seemed a magician, and papers carried praises for his charm, his exuberance, his ability to hold an audience entranced and spellbound. Leibniz once desired a science carnival, but it was Davy who achieved it. From his own words, he shares the joy of public success, the adulation of men (and women), and the applause of crowds. He called it "public communication," and he appears as perhaps the first public engagement specialist science had ever known, the nascent role that would become the most important signifier for science to come. Davy was a showman, a salesman, a suave engineer of the public: "I dream of greatness and utility—I dream of Science restoring to Nature what Luxury, what Civilization have stolen from her," but also, Davy dreamed "of unbounded Applause, Amen!"[57] Volta's discovery made it possible, but it required Humpry Davy to make it shine. Here was the scientist and the dapper gentleman, the captain and the engineer in one. It was Davy who brought charged carbon rods together in the brilliant arc light demonstration, and it was Davy for whom *science* was light itself [Fig. 10]: "the dim and uncertain twilight of discovery, which gave objects false or indefinite appearance, has been succeeded by the steady light of truth."[58] In 1815, Davy would take a wholly new light into the Northumberland mines.

In the dark heart of England's interior, bituminous coal hid away in seams that trickled with water—and sometimes ran with underground rivers. No fiction of palatial mountain mines does justice to the cramped, cold filth, the endless black corridors sloshed with water and caked with coal dust. The vast subterranean veins required constant bailing, the lugging of buckets hand to hand in an uninterrupted chain so that digging could continue. It was to relieve this inherent problem that the first steam engines were built. Thomas Newcomen's "atmospheric engine" used a piston in the Cornwall mines in 1712; heated air, once cooled and condensed, created a powerful vacuum—but Newcomen's engine condensed the steam in the single cylinder, wasting fuel. In 1781, James Watt improved the design by adding a separate condenser so that the main chamber didn't have to be cooled at all; far more efficient, this design required less fuel to remain at a constant temperature.[59] Watt also changed the design to be rotary instead of merely up and down, a detail that would open up a wide world of industrial applications later on, especially with his partner, Matthew Boulton, adding in the rather brilliant idea of selling the design with royalties. Shrewd businessmen, Watt and Boulton cornered the market through patent, and Boulton sealed their popularity through public demonstration (and publicity stunts). His famous brag served as the seed of George Shattuck Morison's claims about manufactured power: "I sell here, sir, what all the world deserves to have—POWER."[60] The first steam-powered loom factory opened in 1790 in Manchester, but in the very earliest days of the nineteenth century, the steam engine's most important job remained its utility in getting at power's *source*. Everything came down to coal. Davy himself connected it to "the necessaries, comforts, and enjoyments of life, but also with the extension of our most important arts, our manufacture, commerce, and natural riches."[61] By means of it, he continued, all the advancements and ingenuity of human labor are heightened and extended, and yet those luxuries came with a cost. Men caught lung maladies, inhaling the dust particulate and shivering in cold day in and out. They lost fingers, and they lost their lives. Most died coughing, others crushed under rock, or even submerged if the water rose to the low ceilings. But the coal seams hid another, secret threat, one that announced itself in earth-shattering explosions.

"Soon after my brother's return from the Continent," writes John Davy in 1836, "he entered upon a new train of inquiry: the investigation of fire-damp."[62] *Fire-damp*. The word has fallen into disuse and conjures conflicting images. What could be both damp and ready to ignite? The *damp* here refers to *dampf*, the German word for vapors.[63] We think today of natural gas smelling of rotten eggs, but that odor has been added by man to track an undetectable enemy. Untreated natural gas is actually nearly odorless. The coal seams held pockets of methane, formed from the same hydrocarbons as the coal, but silent, invisible, and deadly. Humphry Davy's publication on flame offers a simple but clear vision of the possible damages: when accumulated in a shaft or gallery and mixed with oxygenated air, a single spark—and particularly the flame or an open candle—will cause it "to explode, and to destroy, injure, or burn whatever is exposed to its violence."[64] He refrains from giving detailed accounts, as such would "merely [serve] to multiply pictures of death, and of human misery" and because "the phenomena are always of the same kind."[65] Many hundreds of lives were lost to mine explosions, but the most famous of Davy's time occurred at Felling mine (near Newcastle) in 1812. Early in the morning on May 25, a hollow boom sounded, followed by what felt like an earthquake.[66] Debris flew into the air like volcano ash, dusting a mile downwind, and 96 miners lost their lives—the youngest of them two eight-year-old boys.* It was for this reason that Davy had come back to the subject of flame, and light, and how to contain it safely. But instead of starting with the existing coal miners' lamps, as other inventors had done, he started with the *gas*.[67]

Davy's object was to find a light that miners could use safely even in the presence of an explosive atmosphere—but which would *also* consume the "fire-damp" or methane gas. He started at the most minute level; Davy began his career as a chemist, and so he examined the chemical compounds of the gas to work out its several elements: hydrogen, oxygen, and carbon. Through lab work, he realized that it required a great deal of atmospheric air before it would explode, but that "when mixed with three or four times its bulk of air, it burnt quietly."[68] He changed the quantities methodically, tried different thicknesses of glass, and measured the heat of combustion. The most remarkable conclusion had to do with the latter—temperature held the key, and opened the "possibility of constructing a lamp, in which the cooling powers [. . .] should prevent the communication of explosion."[69] Glass caused the temperature of combustion to spike dangerously, but certain metals, and later a fine mesh of iron, could keep the lamp burning without igniting the gas. Davy related the discovery in "cool" terms, as well,

---

* Michael Hunter and Thomas Gordon died with the rest of the men in their family that day; it was common for the youngest to work as "trappers" responsible for opening and closing doors. A great number of the dead were under the age of twenty.

providing a reserved account in print of the trials to reach his conclusions. Letters and journals say otherwise though. Richard Holmes reports that Davy's "true genius as a man of science—his impetuosity, his imagination, his ambition and his seething energy" rushed him toward better models, but he also harassed his associates and assistants—notably Michael Faraday (who would play a much bigger part in the nineteenth-century science to come).[70]

Why the pretended calm? The collected manner in which Davy introduces the lamp doesn't represent a lack of energy and inspiration; it was instead a *calculated* move. The published account reads like a story, moving from Davy's presentation of the problem (the need for coal to fuel England's bright future and the devastation of explosions) to his remarkable solution (the safety light), and never once throws the reader off track. Simple, plain language, presented as a "seemingly inevitable" discovery, removes any possibility that Davy's lamp owed its invention to luck, chance . . . or even Providence.[71] Even Davy's setbacks only confirm success—and we hear the whisper of Victor Frankenstein in those words: "I doubted not that I should ultimately succeed," says Shelley's Romantic scientist. "I prepared myself for a multitude of reverses; my operations might be incessantly baffled, and at last my work be imperfect, yet when I considered the improvement which every day takes place in science and mechanics, I was encouraged to hope my present attempts would at least lay the foundations of future success."[72] Davy works tirelessly, until he at last creates a platinum cage of wire gauze where "fire-damp may be entirely consumed without flame, yielding only a beautiful light."[73] Davy had achieved the impossible, and the "safety lamp" would be hailed as the single greatest public achievement of his life, and upon testing the lamp 1,000 feet below the surface, John Buddle (a mining engineer) spoke as though he had seen a miracle of the first order: "it is impossible for me to express my feelings at the time when I first suspended the lamp in the mine, and saw it red hot [in the explosive mixture]; if it had been a *monster destroyed*, I could not have felt more exaltation than I did" (author's italics).[74] Frankenstein creates a being we have, through countless revisions and adaptations, called a *monster*—Davy gets credit for slaying one. He wasn't without his detractors and rivals for this place in history (and in fact patent disputes turn up even as his lights are being delivered to collieries all through the Northeast), but Davy's coat of arms tells a tale of undampened success: *Igne constricto vita secura*—fire restrained, life is secure. Or, as provided by Holmes from the coat of arms illustration in 1829, *I Built the Light that Brings Safety*.[75]

Davy completed the design for his crest of light in 1817. A year later, Mary Shelley published *Frankenstein*, in which Davy (at least his darker shadow) appears, most particularly in the form of the learned professor and lecturer whose praise of chemistry set the young Victor on his path of destruction. John Abernethy, professor of anatomy at the Royal College of Surgeons, described the power of Davy's lectures in his *Enquiry into the Probability and Rationality of Mr. Hunter's Theory of Life*, 1814. "The experiments of Sir Humphry Davy seem to me to form an important link in the connexion of our knowledge of dead and living matter," he explains. "He has solved the great and long hidden mystery of chemical attraction by shewing that it depends upon the electric properties which the atoms of different species of matter possess."[76] He goes on to say "that electricity is something, I could never doubt [. . .] therefore it follows [. . .] that it enters into the composition of every thing, inanimate or animate."[77] Davy's achievement may have been the creation of safe light, catching, like Franklin before him, an incredible power in a bottle and subduing it. But in the mind of Mary Shelley, these experiments with vitalism and galvanism, the promise that electricity offered life itself would combine with her own experience of death and disorder. She'd lost her mother, who died of fever after giving birth to her. An outcast from her stepmother, she read books atop her mother's grave—where she met the dashing (and married) Percy Shelley. Following him alienated her yet again, and the pregnant and unmarried Mary would lose her first child in 1815; the baby lived only two days. She never gave her a name, and though Shelley would lose more children, the first haunted her nightmares. Victor's creature, also nameless, though far more articulate in the novel than in the early films, calls himself "the miserable and the abandoned [. . .] an abortion, to be spurned at, and kicked, and trampled on."[78] He is forgotten. Ambiguous. The "light" that broke in upon Victor Frankenstein means to be a scientific one, not the flash of the magician, but his creature remains as horrifying as a conjurer's monster. Victor Frankenstein is not meant to be a supernatural villain; he is instead a genius, ready to risk consequence for discovery, to proceed at any cost to himself or to others, a willing acolyte to the progress of science. And perhaps, suggests the novel, this is worse. It's death, not life, that Victor wields. Put another way, his science released dread tech into the world rather than serving as protection from it. In 2012, Shelley's novel was rereleased with steampunk illustrations from Zdenko Basic and Manuel Sumberac, full of twisted metal and gearwork that turns organic creation to automaton AI. Frankenstein's "monster" appears in the steampunk *Van Helsing*, too, and though the movie makes no bones about including supernatural vampires and werewolves, the creature has been *built* not born. Instead of the uncanny dread of the "mother machine," here we have the motherless being, a thing born of electric fire engineered by man—and full of man's desires. Davy fought the darkness by being a worshipper of light, but even he took his risks in the dark.

Unwell and alone in Rome, Davy wrote back to his brother in consideration of all that had been lost when science banished superstition, when chemicals and batteries took the place of divine visions and poetic expression. Then he recounts a kind of near-hallucination: "I had on the 7th April, 1821, a very curious dream," the letter begins. "I imagined myself in a place partially illuminated with a reddish light; within it was dark and obscure; but without, and opening upon the sky, very bright."[79] At the division of dark and light, Davy "experienced a new kind of sensation [. . .] as if I became diffused in the atmosphere." He rises upon wings, to find himself in the same dread cosmos of Newton and his forebears: "I, for some time, reposed upon the highest of these galaxies, and saw as it were the immensity of space—systems of suns and worlds, forming a sort of abyss of light."[80] In this transfixed state, Davy exclaims, "I had always been of opinion that the spirit is eternal, and in a state of progression from one existence to another more perfect; that I had just left a world where all was dark, cold, gross, and heavy; that I now knew what it was to have a purer and better existence."[81] The dialogues he wrote at the end of his life continue in the metaphysical vein; he ached for a world of "intellectual light," where the causes of all things would be understood, and where the ultimate pleasure would be "unbounded knowledge." The future would tell a story where darkness retakes the battlements, smudging it with war and industry—some of it made possible by Davy's own safety light, as men burrowed out coal for boilers, factories, and a growing and greedy population. He wanted Science to champion Nature, but in the end, it mostly pillaged. Davy's career ended in frustrating turmoil as he fought with other intellectuals, even Faraday, whom he once championed—and his marriage collapsed into increasing disharmony. He died of stroke in Geneva in 1829, his last work finished just before (and in which his metaphysical dialogues appear): *Consolations of Travel, or The Last Days of a Philosopher.* His brother's hope was that Davy might, himself, be a "beacon of light to young and erring genius,"[82] in the way that Frankenstein hoped, by his example to "Learn from me, if not by my precepts, at least by my example, how dangerous is the acquirement of knowledge and how much happier that man is who believes his native town to be the world, than he who aspires to become greater than his nature will allow."[83] Instead, Davy's strange last farewell, with its half-strangled and desperate desire for light, its fictional flights and disembodied cosmic dialogues, would "take a surprising hold" on the next generation of scientists to come, offering a broad vista and even a map for pursuing nature to her hiding places, come what may.[84] The urge for knowledge and for light kicked off a new generation of discovery, not only for the chemist and natural philosopher, but for the explorer willing to follow it into dark country.

# FOUR

## Into Dark Country

magine a landscape of heaving white pinnacles, where darts of sheer crystal pierce skyward, shatter, and reform in spiked fortresses of ice. Liquid water swells through broken holes as black as night sky; the sun never rises, but slides around earth's rim, ever distant, far, and white. The Arctic induced snow blindness; in desperation did men search for color, for contrast, and in desperation listened to the howl of winds that obliterated senses. We cannot now fathom the hope that drove explorers north: that when they reached the pole, they would find a paradise instead of frozen tundra, a passage, a land of wonder. "I try in vain to be persuaded that the pole is the seat of frost and desolation," says Walton in the first pages of *Frankenstein*. But despite the wracking cold all around him, he believed that where the sun is ever visible, it must produce a glorious country of beauty and eternal light. Walton's fervor echoes 1 John in the King James Bible: men should seek God, for "God is light, and in him is no darkness at all," except that Walton instead seeks a natural deity, "the wondrous power which attracts the needle,"[1] and the establishment of his own legacy: to *see* what no man has seen, to walk where no man has walked. The explorer does not mean only to commune, but to command, to see but also to *understand*. Newton wanted to order the cosmos; explorers wanted to order their world.

The great paintings depict stalwart captains, standing upon decks and decrying fate for fortune, defying loneliness and trauma and death, and seeing in the wasteland endless possibility. But there's more to this story than captains. The cramped cabins also housed botanists, physicians, collectors, and cartographers: men like Hans Sloane, whose specimens form the basis of the British Museum; Sir Joseph Banks, long-standing president of the Royal Society; and Mungo Park, Scottish botanist and the first westerner to track the vastness of the Niger. The explorers included in their ranks the ill-fated Robert Falcon Scott, who, like Walton, sought the land with no night and paid dearly—and a young Charles Darwin. Money, power, and politics might fund supplies and stock ships, but they could not inspire. An explorer's heart strikes sparks from wanderlust, curiosity, and a blazing sense of the great, wide, and glorious world. It burned in the breast of Jules Verne, too, and though he never piloted a vessel of his own, he would steer an entire genre of extraordinary journeys.

The Unknown was an invitation; like Davy's last dialogues it scintillated in its ambiguity. The explorers did not *fight* the darkness; they raced to its embrace. "If human beings are attracted to the known, to the realm of things as they are," writes Elting Morison, "they are also, regrettably for their peace of mind, incessantly attracted to the unknown and things as they might be."[2] The story of electricity and the story of exploration together fired the imagination of generations to come—they inspired some of our greatest fiction, gave rise to heroes and villains before unimaginable. But most importantly, they serve as the constituent parts of George Shattuck Morison's *New Epoch*. The heroes of electricity first manufactured power—that wondrous ability that Morison claims will sweep all else before it. "Man's capacity is measured by the power which it can control," a power that has "determined the ability of one tribe or nation to rule another."[3] But the explorers, by their ships, their maps, and the soul-stirring power of their stories, would manufacture an *empire* on which to practice power. Men would die (usually horribly) making inroads to new worlds. And men would die when they succeeded. The steampunk ethos, the fictional worlds of *Whitechapel Gods* or *Frankenstein* or myriad other tales help elucidate the strange underbelly of scientific desire and dread. But every explorer's tale is also, by its very nature, a fiction. And here we see the great change at its beginning: We can follow the careful experiments of the electricians ourselves, and by those methods re-create wonders, but the voyages into dark continents, down jungle river valleys, across frozen poles, can never be replicated. The only witness also tells the tale, and spins it often years and years later, when age and illness have rendered them incapable of returning to the wild, when their journeys have become idyllic, even to themselves. Or, alternatively, the story is turned out in haste and to the political taste of whomever it serves and whoever will pay for the next voyage (and without the codicil that explains where, in the end, the explorer's body lies). The success of the story did not depend on its veracity, but on its variety. Science and fiction begin their ages-long dance, and by the end of the nineteenth century, scientists, engineers, and explorers will become fictionalized heroes (and at least one fictional detective will seemingly become fact).

In 1596, Sir Walter Raleigh reported "headless men" lived in the country of Guiana, and that oysters grew on the mangrove trees. He claimed the natives lived in golden cities, that they bathed in gold dust, and that the land's riches were open to any who would possess them. Even the title of his publication serves as advertisement: *The Discovery of the Large, Rich, and Beautiful Empire of Guiana*, and his reason for writing was to encourage England to possess it. Historian Willard Wallace once called it a "treatise of empire," and Raleigh a colonist before his time.[4] The early days of exploration, aided as they were by the astronomers of the sixteenth and seventeenth centuries, looked at new worlds as lands of enrichment. The discovery of the Americas resulted in a race for consumption—the Spaniards leading— as they mauled the land and people for all they could carry. We rightly remember conquistadors for their conquests, epidemics, destruction of existing systems of government, and the razing of the Aztec, Inca, and other empires. Enslavement and genocide—some estimates suggest that before Cortez arrived in 1519, 25.3 million people lived in the discoverable parts of the New World. By 1605, only one million remained.[5] More conservative estimates still point to mass death and destruction, a legacy of horror that we can never forget. By the eighteenth century, the desire for expansion had become an empire-building strategy—not to strip the land but to husband it, not to destroy the population but to colonize it. The results, not infrequently, were the same. But the meaning, and so the characters, of the new program had changed.

When Hans Sloane made it to Jamaica, the world already knew of its existence and had already heard tales of its inhabitants. It had, in fact, already been colonized. Here would be no tales of one-legged men, or people whose mouths appeared in their stomachs, or of mermaids, mermen, and monsters. Sloane was an Irish doctor, personal physician to the island's governor. He was also a botanist with a mad desire for collecting. Lizards of all shapes and sizes, hundreds of birds and insects and plants—all of them utterly new and strange—made their way back to the curious in England. He wrote about the utility of the land, about the possibilities of the new plant specimens for medicine, and even about the witch-doctors he encountered there, who spoke of obeah—or *Voodoo*. The slaves, it was said, could practice this medicine against their masters.

A white Jamaican planter, Edward Long, described the ritual: "When assembled for the purposes of conspiracy, the obeiah-man, after various ceremonies, draws a little blood from every one present; this is mixed in a bowl with gunpowder and grave dirt; the fetish or oath is administered by which they solemnly pledge themselves to inviolable secrecy, fidelity to their chiefs, and to wage perpetual war against their enemies; as a ratification of their sincerity, each person takes a cup of the mixture, and this finishes the solemn rite."[6] Obeah/Voodoo originates in the Caribbean, and thrives on the belief that spirits of the dead can be harnessed to attack the living. It also relied upon the use of herbs—and of poisons. It was with the poisons that Sloane chiefly concerned himself; European doctors had much to gain from the promise of plants, and also as much to lose. Obeah men, as centers of the community, could promise solidarity and resistance; they could threaten with invisible power. Tessa Harris's novel *The Lazarus Curse*, set in the eighteenth century, plays with the power and resistance theme by transporting Voodoo back to England in a story that revolves around the power of chemicals: the still-undiscovered alkaloids hiding in vegetable matter, from opium to nightshade. Tales circulated then as well as they do now, and in 1816, a law was passed recognizing the danger of Obeah: "if there shall be found in the possession of any slave any poisonous drugs, pounded glass, parrot's beaks, dog's teeth, alligator's teeth, or other materials notoriously used in the practice of Obeah or witchcraft, such slave upon conviction, shall be liable to suffer."[7] The white owners maintained at once that the black slave had no power—and yet lived in fear of obeah magic; back in England and unable to parse the fact and the fiction, eager minds wondered at what other wonders might still be discovered. Among them, Joseph Banks, the same man who would, in his later years, promote both the stargazer William Herschel and the inexhaustible Humphry Davy. And, years later, the eager mind of a young boy at a boarding school in Nantes, France: Jules Verne.

## Terra Icognita

Jules Verne was born in 1828, in the same year that French abolitionists finally pushed the government to end the trade and trafficking of slaves—though it would continue covertly until Verne was a young adult of twenty. Nantes, the town of Verne's birth, served as the launch point for slaver ships, and so remained the focus of abolitionists and government interference, a foment that would stay with Verne all his life (and appear in works like *20,000 Leagues Under the Sea* and *The Steam House*). The trade, the town, the port, for which Nantes is well known, alone would instill a child with wonder at the wide world beyond—each day a new face and new discoveries. He read widely, too, of the explorers themselves, great men with great dreams and great worlds to traverse. But if the narratives frequently offered a tissue of errors (as Verne himself complains later), his own mythos leaps and bubbles with excited, if not entirely verified, stories of adventure. The first "authoritative" biography appeared in 1928 by Verne's grandniece, and told the tale of his boarding school experience, where the widowed teacher spun tales of her shipwrecked husband—lost for thirty years. It also recounts his flight from home, wherein the young Verne stowed away as a cabin boy on the *Coralie* freighter.[8] While most of these stories are specious, one thing remains vibrantly apparent: Verne had a great love for scientific discovery and an equally unassailable affection for literature. In a letter to his father while studying law, he explains, "Literature above all . . . my mind is focused uniquely on this goal! What's the use repeating all my ideas on the subject [. . .] my dear father, whether I do law for a couple of years or not, if both careers are pursued simultaneously, sooner or later one of them will destroy the other"—and, Verne patiently explained, "the bar [law] would not survive."[9] Verne's earliest articles, published for Pitre-Chevalier, captured the spirit of discovery: "The First Ships of the Mexican Navy," "A Balloon Trip"—and, notably—"Wintering in the Ice." The love of voyage never left him, and by 1880, Verne completed a history of the eighteenth-century explorers who so fired his imagination; nearly four hundred pages, twenty engravings and maps, and his own pronouncement that all great expeditions owe their allegiance to "immense progress" of *science*.[10] In the years between, Verne first conceived of his *roman de la science*, the novel of scientific significance from which Nemo's *Nautilus* would later emerge. And the rest, in a very literal sense, is history.

The adventurous *20,000 Leagues Under the Sea* gives us gadgetry through the electric and gear powered *Nautilus*, and *The Steam House* features a mechanical elephant trundling through the Indian jungle. But even here, there's more to mimicry than clockwork conchs and steam-powered pachyderms. Captain Nemo uses his ship to sever all ties with humanity following the devastatingly bloody "Indian Mutiny" of 1857. The steam elephant serves as a safety vehicle for British colonists (whom Nemo calls the "oppressors") in the aftermath of the same rebellion. *Eight Hundred Leagues on the Amazon* and *The Mighty Orinoco*, and even *The Mysterious Island* feature landscapes awash in rebellion, in bloody battles and the deaths (or near-deaths) of those who seek to explore and profit. Verne's nonfiction account of the explorers whose shadows he chases in fiction begins, likewise, with needless bloodshed practiced by Dutch explorer Roggeveen, who fired upon "an inoffensive population which had awaited them upon the shore, and whose only fault consisted in their numbers."[11] We remember the bright adventures. But we forget the darkness, says Verne, at our peril.

Roggeveen's company played a dangerous game. Like all eighteenth-century navigators, they fumbled in the dark to find islands whose coordinates were ambiguous, or even the stuff of imagination and legend. Food ran short, or went sour, and men suffered horribly with scurvy, crawling into their hammocks sometimes never to rise again. When they came ashore, it was as desperate men, but they pressed their advantage in weapons and massacres "worthy of barbarians" rather than "civilized men."[12] The British Lord of the Admiralty couched these exploits in patriotic terms: "nothing contributes more to the glory of this nation, in its character and of a maritime power, to the dignity of the British crown, and to the progress of national commerce" than exploration. Go, therefore, and bring light to the unknown places. Commodore Byron received that office, and took command of *The Dauphin*, a man-of-war, in 1764. The ship went heavily armed: 24 guns, 150 sailors, 37 petty officers. Byron landed at Rio to aid the sick, and was only then told of his secret mission—to hunt out islands in the Southern Seas, for England. In those climes, he encounters the extraordinary: an island, where waves lapped a pleasant shoreline and two steep mountains rose from the water. The vessel chased the island for hours, only to discover a mirage: "only a land of fog! I have passed all my life at sea," Byron's log explains, "but I never could have conceived so complete and sustained an illusion."[13] On other islands, he encounters men "of gigantic stature, and looking like monsters in human shape" with bodies "painted in a most hideous manner."[14] Byron's crew submitted to extraordinary heat, to privation, and to landings which continued to end in the deaths of natives, but he returned to England without having laid claim to new discoveries. His ship, however, would carry on—and under the captaincy of a man called Wallis, she would land in paradise.

So many of the islands offered only desolation and "resembled the ruins of the world" rather than "an abode of living creatures."[15] Wallis's men suffered constant damp, a rain that would not permit even one dry corner (all the foodstuffs were spoiled rotten), and were afflicted with fever and ague, distemper of bodies and minds, and long hours under sail and in baking sun. Coleridge's *The Rime of the Ancient Mariner* of 1834, built from the stories of seafarers and whalers, reimagines the sick want of breeze and freshening air:

> All in a hot and copper sky,
> The bloody Sun, at noon,
> Right up above the mast did stand,
> No bigger than the Moon.
>
> Day after day, day after day,
> We stuck, nor breath nor motion;
> As idle as a painted ship
> Upon a painted ocean.

The same might be said of the fetid water, still and turning bitter to the stomach and worse to the intestines: "Water, water, every where,/Nor any drop to drink." The *Dauphin* skirted land it could not touch, turned away by hostile natives or sheer cliffs. Imagine, then, the delight of sailors so long at sea when the painted sky opened over a wondrous jungle. "The huts of natives were sheltered by shady woods," the account recalls, with "graceful clusters of cocoanut-trees," and wooded summits, and best of all "the silver sheen of rivers glistening amid the verdure as they found their way to sea."[16] The natives invited them to shore, especially the women (so the narrative goes) with "unequivocal gestures," and the men brought fruit and fowls and pigs as presents. The captain had discovered his paradise, a promised land, and one that Captain Cook would return to with the best of botanists, Joseph Banks. But for all the talk of Eden, the natives called it *Tahiti*.

Richard Holmes's account of Banks's journey begins as follows: "On 13 April 1769, young Joseph Banks, official botanist to HM Bark *Endeavour*, first clapped eyes on the island of Tahiti, 17 degrees South, 149 degrees West. He had been told that this was the location of Paradise: a wonderful idea, although he did not quite believe it."[17] He'd joined Captain Cook's expedition to draw plants, collect specimens, and provide what Wallis's trip had not: scientific observation—but he, not unlike Humphry Davy after him, had a poetic turn. Despite initial misgivings, Banks described this place as "the truest picture of an Arcadia of which we were going to be Kings that the imagination can form."[18] Like the men of the *Dauphin*, the sailors found the islanders welcoming, and the women "free" with their affections. A sexual trade between the sailors and the inhabitants (in which sex could be exchanged for any metal item) had rendered Wallis's ship endangered—the crew pried up the boards to obtain any screws, nails, bolts, and iron that united the timbers, and when they returned to sea, a storm carried away the *Dauphin*'s poop deck and forecastle.[19] Cook tried to be more careful of his *Endeavour*; they were there, after all, to see the transit of Venus across the face of the sun and to establish the solar parallax.[20] But Venus was tempting Banks in her other guise; he spent enough time on the island to take a "wife," the lovely servant of the Tahitian queen, to learn their language, and in many other respects behave and live as the natives lived. He was, says Holmes, pioneering a different kind of science—what we think of today as anthropology—and unlike the bloody voyages before him, he "trusted [himself] among them almost as freely as I could do in my own country, sleeping continually in their houses in the woods with not so much as a single [white/British] companion."[21] He had gone, with the fervor of Enlightenment burning in his breast, to practice "pure" science; to elucidate dark corners with the light of new mechanism. A colleague listed the materials that went into the hold: "a fine library of Natural History [. . .] all sorts of machines for catching and preserving insects; all kinds of nets, trawls, drags and hooks [. . .] they even have a curious contrivance of a telescope by which, put in water, you can see the bottom at a great depth."[22] The ship had been previously known as the *Earl of Pembroke*, but the British navy purchased her for exploration—to hunt the *Terra Australis Incognita*, the "unknown southern land."[23] They would also witness an astrological event that would not take place again for one hundred years. Light, or at least enlightenment, would carry them into parts unknown, and all under the flag of a now *international* science, a public project followed as closely from home. *The London Gazetteer* reported shortly after they sailed: "[t]he gentlemen, who are to sail in a few days for George's Land," or the newly discovered Tahiti, "with the intention to observe the Transit of Venus, as likewise, we are credibly informed, to attempt some new discoveries in the vast unknown tract" of the Southern Seas.[24] They did not find them. For Banks, "some pleasure" could come of finding nothing at all, if only to disprove "that which does not exist but in the opinions of Theoretical writers," men wrote about the seas without ever being in them.[25] Banks, in his role as jungle explorer and later his role as head of the Royal Society, lived in a world far removed from that of Newton. For Banks, only seeing was believing, and though they might be looking at the same constellations as their predecessors, for captains like Cook (and in fact the British crown and navy) those stars were useful mainly as a means of recording coordinates, and of expanding and enriching the empire.

For Jules Verne, Captain Cook remained the most singular of explorers, "the most illustrious navigator whom England [can] boast."[26] Nearly two thirds of his book on eighteenth-century navigators is dedicated to Cook's three voyages, and his rise from humble beginnings to greatness. Cook undertook the Tahitian voyage at the age of forty and, well-warned by Wallis's predicaments, gave strict orders for his men to avoid injuring the natives. But Banks's journals tell of repeated and painful incidents, where common thieving met with severe punishments—and in almost every incident, the natives were killed by gunfire. Verne recounts how Cook "took possession" of islands, of boats, or harbors—of how they moved on from Tahiti to New Zealand, and the bloodshed that followed. Cook complained that, surely, "human and sensible people will blame me for having fired upon these unfortunate Indians," who "did not deserve death for refusing to trust."[27] And yet, he exonerated himself and his men using the same language of discovery found in the rhetoric of the experimenters, who toyed with electricity despite danger, concocted gases and explosives without consideration of consequence, and (in the case of the fictional Frankenstein and the actual Aldini) attempted reanimation of corpses without scruple. "My commission by its nature," Cook explains in his journals, "obliged me [. . .] [and] I could only do so by penetrating into the interior" even by means "of open force."[28] To throw the doors off of nature, to bring the light, to map and to understand outweighed other considerations. If they could not be bought with trinkets, they would be stymied by force, their land and provisions taken by those who claimed their islands for the crown. Cook's first voyage earned him the rank of commander, and he set out for a second—the jungles of Tahiti, even the lands of Australia and New Zealand, did not answer his duty. Cook set out to find the mystery land of the southern pole.

## *In Ice and Snow*

*"I am going to unexplored regions, to 'the land of mist and snow,' but I shall kill no albatross; therefore do not be alarmed for my safety or if I should come back to you as worn and woeful as the 'Ancient Mariner.' You will smile at my allusion, but I will disclose a secret. I have often attributed my attachment to, my passionate enthusiasm for, the dangerous mysteries of ocean to that produc-tion of the most imaginative of modern poets. There is something at work in my soul which I do not understand."*

—Walton's second letter, Mary Shelley, *Frankenstein*

Walton, intrepid explorer of Shelley's *Frankenstein*, has a secret. He has pitched his career, a dangerous cycle of unsafe seas and unknown harbors, on the hopes of a narrative *poem*. Samuel Taylor Coleridge, a "Lake poet" and friend of Wordsworth, a man with dark demons, but whose brighter lights strongly influenced a young Humphry Davy, composed *Rime of the Ancient Mariner* about a ship lost beyond the 40th degree, somewhere in the Antarctic Circle. The ship ultimately ends under the hot eye of an unforgiving sun, but it starts "through the drifts [and] snowy clifts" where "ice was all between."[29] The ship's predicament falls to the bad faith of the Mariner, who kills an albatross and later must pay for his treachery by losing his captain and crew and hosting supernatural beings. Walton's own ship is likewise locked by "vast and irregular plains of ice," which seem to have no end—and though he promises to avoid the fate of Coleridge's hero, he nevertheless has "something at work" in his soul that he can neither under-stand nor master. Unlike Captain Cook, Walton has nothing pushing him to the nether regions of the world except his "enthusiasm" for "dangerous mysteries." But even Cook pursued his mission beyond the hopes and cer-tainly beyond the worst fears of his own men.

Today, we know that Antarctica, a continent twice the size of the conti-nental United States, sits at the southern pole. But in the eighteenth century, the assumption that land must be found in the south arose from no scientific knowledge at all, just assumptions about *balance*. If the globe had so many continents to the north, they reasoned that just as many must lay to the south; the idea was present even in the work of Ptolemy, and as empires expanded through the centuries, more ships went in search of them. British explorer Matthew Flinders would ultimately claim Australia as the promised terra incognita in the early nineteenth century, writing with certainty that nothing could exist below. The *Endeavour* never ventured so far south, though Cook's ships *Resolution* and *Adventure* crossed the Antarctic Circle in the 1770s, coming within about 75 miles of the coast before retreating from crushing sea ice.* For weeks together, Cook and his men saw great bergs—large as land masses and treacherously shedding themselves into the sea with earsplitting cracks and groaning sighs. The danger increased as they moved further south, until the field of ice extended beyond the visible horizon. Cook did not change course, but continued on steadily south. Birds visited Cook, too, not only albatross but penguins and other sea birds, but—says Cook's journal—"we had now been so often deceived by these birds, that we could no longer look upon them [. . .] as sure signs of the vicinity of land."[30] They had signs of other kinds, however, not so violent as Frankenstein's lightning strike but at least as awesome in wonder. The seas had calmed, the night cleared. And there, between midnight and three in the morning, a light appeared in the heavens. "It sometimes broke out in spiral rays," Cook explains, "and in a circular form [. . .] its light was very strong, and its appearance beautiful [. . .] it diffused its light throughout the whole atmosphere."[31] Knowing already of the northern lights, or aurora borealis, Cook calls the phenomena *Aurora Australis*, after the unknown world of the south that he continued to seek. "It was seen first in the east, a little above the horizon; and in a short time, spread over the whole heavens": light, even in the darkest corner of the world, though winter was closing

---

* The first documented landing wouldn't happen until 1895.

in white on white, deadly and solid—and night would close them up, too, long and lasting.

What drives humankind to the edges of experience? Cook's journey presages fiction, but also inspires it: Walton's journey in *Frankenstein*—Coleridge's *Rime of the Ancient Mariner*—and later, Edgar Allan Poe's *The Narrative of Arthur Gordon Pym* give us pole-hunters, explorers who leave the habitable regions and look for new worlds beyond the ice. Cook may have been searching for the southern continent in the late 1700s, but in the early 1800s, nations vied for discovery of the *northwest passage*, a means of crossing from the top of the world to its nether side. Gothic novelist Wilkie Collins writes *The Frozen Deep* about a lost explorer (capitalizing on a true and unfolding tale about the ill-fated voyage of Sir John Franklin) [Fig. 11]. The perilous journey and its unyielding frozen wasteland drew imaginations, not for what it contained but what it promised. What if, as Walton believed, the force that drew the needle north was a life-giving force? What if a paradise akin to Tahiti's tropic shade awaited? The notion seems foolish to the modern reader, but the ideas that fueled fictions like *Pym* or Verne's *Journey to the Center of the Earth* evolved from scientific minds working over scientific problems. Edmund Halley, rough contemporary to Isaac Newton, promoted just such an unusual notion. It took account of steam vents and volcanoes, of rotation and magnetism, and suggested that the earth must be *hollow*.[32] Nested spheres, each with its own ecosystem, might be entered, he and his followers surmised, through the northernmost pole—something Verne would, once again, capitalize on. Two stories emerge: one of an undiscovered continent, and one of an undiscovered polar world. The first, despite Joseph Banks's misgivings, would turn out to be true; the second, false—but the race for the North Pole would carry on anyway, in one long and devastating journey after another.

The arctic, for humankind, remains one of the world's most inhospitable places. But its blank, white desolation offers, in its way, the counter to unremitting darkness. Both represent an unknown that, in the age of exploration, did not frighten—but rather, invited. Victor Frankenstein speaks of darkness and light in inverse terms; ignorance offered safety and comfort, enlightenment resulted in tortures and destruction. In Poe's *Pym*, say critics Frank Frederick and Diane Long Hoeveler, "blackness" corresponds to "the dread of mechanical time" while "whiteness" results in a numbness and illusion that is both preferable and totally annihilating.[33] And to cite a different tale, which is as "steampunked" as *Frankenstein* (most notably in China Miéville's *Railsea*), Herman Melville's eminently inspirational *Moby-Dick* describes the perilous desire and dread for "whiteness." It's very "indefiniteness" shadows forth "the heartless voids and immensities of the universe, and thus stabs us from behind with the thought of annihilation, when beholding the white depths of the milky way"—explains the narrator, alighting upon the foreboding of seventeenth-century chaos. "Or is it, that as in essence whiteness is not so much a colour as the visible absence of colour; and at the same time the concrete of all colours"—the very theory that Newton himself made possible. For these reasons, Melville writes, that a wide landscape of snows thrilled and terrified, "a colourless, all-colour of atheism from which we shrink."[34] I have written that all explorers' tales were, in their way, selling fictions; likewise, the adventure novel sells with truth. Poe borrows heavily from the American explorer Jeremiah Reynolds, and Melville from tales of the vessel-destroying whale Mocha Dick. The legend and the spirit of adventure, the mythmaking of a generation, and the magnetic draw not of the poles but of the great unknown, lingers on.

A hundred years after the publication of *Frankenstein*, search parties would discover the remains and grisly diary of Robert Falcon Scott, an arctic explorer who had, by November 1912, been dead and frozen for eight months; they had been in a race against a team of Norwegians to be the first to discover the pole—and failed. "Great God! this is an awful place," writes Scott in what would be his last testament to the living world, "and terrible enough for us to have laboured to it without the reward of priority."[35] Famously, Scott's companion Titus Oates, knowing he was too weak to go on, walked into a blizzard to save his companions the trouble of caring for him—"I am just going out," he said, "and may be some time." They died only eleven miles from the depot that could have saved them, and Scott's last entry reads, pitifully, "For God's sake, look after our people."[36] Author, curator, and explorer James P. Delgado tells of the strange remains still to be found scattered across the north, "fragile and yet strangely resilient [. . .] be they a ring of rocks that once marked a tent, or the intact hulk of *Breadelbane*, crushed and sunk in 1853."[37] Such relics still inspire stories, such as the steampunk graphic novel *The Arctic Marauder*, a northern recasting of *20,000 Leagues Under the Sea*, wherein the eccentric scientist Louis-Ferdinand Chapoutier pilots an iceberg ship to disperse chemical weapons. *Artic Marauder*, says steampunk critic Erika Behrisch Elce, "overlays the romance of Victorian scientific exploration with a modern tale of cruelty and violence" and "scientific hubris."[38] But as the narratives of Banks, Cook, and, later, Jules Verne (who is the namesake for the hero's original sailing vessel) attest, the combination of exploration and cruelty isn't modern at all. Captains and sailors assailed the islands with weapons and disease, and explorers died of fever, dysentery, starvation, scurvy—or froze to death, drowned in frigid seas, or were carried off by hungry animals of the north and far south. Cook himself died at the hands of natives on his third voyage, with his crew bartering with the Hawaiian natives to get back his head and hands (the only parts of him that remained). Delgado, in his twentieth-century trips to the frozen north, describes the process as a spiritual journey: "I have stood on grey gravel beaches, with the wind never ceasing," he writes, "floated in frigid waters [and . . .] mused, alone for an hour, in the small cabin of the *Gjøa.** [. . .] For I . . . sought the Northwest Passage, and found there but the way back home again."[39] For the explorers of ice and snow, however, the white empty void, its howl of wind and crack of ice, is all the home there is, a great white casket entombing dry bones. And still, Sir Joseph Banks spent his last years pining that he never again went to sea.

---

\* The first vessel to make transit of the Northwest Passage.

## Dark Continents

Holmes recounts the decline of Sir Joseph Banks in his waning years; crippled by recurrent gout and swollen legs, both literally and figuratively bound to his presidential chair at the Royal Society, Banks's dreams and desires stir further and further afield. "From Soho Square," writes Holmes, "[Banks's] gaze swept steadily round the globe like some vast, enquiring lighthouse beam."[40] He could no longer make the journey, but he could vicariously navigate the perilous beyond through the exploration and writing of others. In 1803, Banks reminds his reader of the aim of all science—to peer, to penetrate, to risk: "I cannot agree with those who think it too hazardous to be attempted: it is by similar hazards of human life alone that we can hope to penetrate the obscurity of the internal face of *Africa*" [author's italics].[41] The great unknown, unmarked, unmapped territory would be known in years to come as the *Dark Continent*.

American journalist Henry Morton Stanley, the man who hunted down Dr. David Livingstone in Africa in 1869,* typically achieves credit for first using the term "dark continent," but Banks's own words already describe the land as obscure—a blank on the existing maps, with a rough outline penciled in by sailors but without the geographical features, lands, and kingdoms within. That does not mean no one had arrived on those unknown shores; the Greeks and Romans (closer, geographically) settled parts of North Africa, and the Egyptian kingdoms were well known even to Europeans. The Dutch colonized parts of Africa to serve the East India Company, a trading route known for economic empire and brutal tactics, and other nations (including France and England) set up outposts for use by the slave trade. Fort settlements made no attempt at further exploration, but concentrated upon the coasts and the grim holds of slave ships; what more could be had from a land so devastating? Frank McLynn, an early

---

\* Famed for the saying "Dr. Livingstone, I presume?" when locating the missionary in an African village.

biographer of Morton Stanley, described a hellish landscape of heat and disease, where "screws worked loose from boxes, horn handles dropped off instruments, combs split into fine laminae and the lead fell out of pencils . . . hair ceased to grow and nails became as brittle as glass [and even] flour lost more than eight per cent of its weight."[42] Who, he might ask, would desire more of such a land as this? Walton, possibly. "I am practically industrious—painstaking, a workman to execute with perseverance and labour," claimed *Frankenstein*'s erstwhile explorer, "but besides this there is a love for the marvellous, a belief in the marvellous, intertwined in all my projects, which hurries me out of the common pathways of men, even to the wild sea and unvisited regions I am about to explore. [. . .] Shall I meet you again, after having traversed immense seas, and returned by the most southern cape of Africa?" The slavers worked only to harvest men, to barter them, and to brutalize them. The explorers had their sights set upon something else. Sir Joseph Banks founded the Association for Promoting the Discovery of the Inland Districts of Africa, which (despite being abysmally named) set about opening routes through Egypt and the Horn of Africa for discovery, not conquest, enlightenment, nor the trade of souls.[43] They would search for the legendary city of Timbuctoo, a mythic place not unlike the cities of gold sought by the conquistadors, "glittering with towers and palaces roofed in gold."[44] It sat upon the Niger, they claimed, at the convergence of the Arab and African routes. What else might flow down from such ample ports? Science may have been the principal aim, but so, too, was commerce, and where enrichment might be had, politics would follow. None of this would be set out in the advertisements—none of the potential darkness and dread laid bare. Only the bright edge of bright-eyed wishes, the desire to see a new land and bring the light to its dark corners. Hopes were high. But not one of the early explorers returned alive.

Jules Verne recounts the stories of Friedrich Hornemann, who explored Fezzan (modern Libya) and north of the Sahara, Major Houghton in Gambia, W. G. Browne in Darfur, and Mungo Park (supported in earnest by Joseph Banks) as he searched for the Niger and Timbuctoo. Hornemann died in Nupe in 1819, Houghton was betrayed by his guides and perished, and Browne was murdered while trying to reach Tehran. Others succumbed to fevers or disease, like Cook's sailors in Tahiti—others died by accident. But many, many more simply vanished in the dark heart of unknown territory, answering letters with silence, lying in unmarked graves or lost in desert and river, scattered by carrion birds. And still, in 1795, Mungo Park left Portsmouth to follow in Houghton's footsteps. His arrival in Bondou (part of Senegal in West Africa) proceeds without accident; his journal bubbles with enthusiasm and humor. He recounts how the wives of the monarch there prod him about his looks, insisting that his white color and features must be artificial, "the first, they said, was produced when I was an infant by dipping me in milk, and they insisted that my nose had been pinched every day till it had acquired its present unsightly and unnatural conformation."[45] Park recounts kindly and curious treatment throughout, that natives repeatedly counted his toes and fingers to determine whether so odd a creature would be human. He tells, too, of the king of Kasson, who tried to protect him from traveling through war-torn territory. The halcyon days over, Park moves ahead into darkness, into captivity with the Moors of Jarra, his long and dismal trek through monsoon-ridden valleys, his reduction to poverty and nakedness by robbers, and finally, a deepening despair. Alone, hungry, and ill, Park wanders in twilight until a Moorish woman finds him and leads him to a tent. By lamp, she and her companions sing to him, calling him the pitiable, motherless outcast.[46] All we expect from the bold explorer—from the Waltons on their forecastle decks, facing the great beyond—reduces to this, the weary, grief-stricken Park, like Victor Frankenstein on his iceberg. Holmes describes Park's near collapse in romantic and poetic terms; like Coleridge's *Ancient Mariner*, Park's despair is overturned not by Providence and prayer alone, but by nature—he recognizes a tiny plant, the size of a thumbnail, blooming underfoot. His scientific curiosity reengages, and (much more like the mythic hero) he starts up, banishes thoughts of death, and carries on. Park returns to England and publishes his travels, which influence more than one writer to be. Not only Jules Verne, who would go on to write about jungles and exploration of unknown parts in his *Voyages Extraordinaires*, but also Joseph Conrad: "In the worlds of mentality and imagination which I was entering, it was they, the explorers, and not the characters of famous fiction who were my first friends."[47] But Mungo Park, or at the least, the Mungo Park who appeared in the pages of *Travels in the Interior of Africa*, was a fiction of the first order. Holmes calls Park, himself, the terra incognita, and why not? Who were these men, but the stories they presented? And who but these men, or others who would risk all to follow them, could ever contradict the story they revealed? Fact and fiction merge, and converge, and the public turned selected stories into bestsellers, lighting up the grimmest part of each tale with a golden and iridescent light. The great race for Africa had begun, a "scramble" of invasion, division, colonization, and imperialism that began in the Victorian era and extended to the first whisper of World War I. And in the wake of devastation wrought by the new century, all that is "light" about Enlightenment exploration seems very dark indeed.

Joseph Conrad published *Heart of Darkness* in 1899. The novella begins with the same exuberant descriptions that appear in Sir Joseph Banks's and Cook's accounts of Tahiti, and in Walton's excited pronouncements about paradise: "Watching a coast as it slips by the ship is like thinking about an enigma. There it is before you—smiling, frowning, inviting, grand, mean, insipid, or savage, and always mute with an air of whispering, 'Come and find out.' This one was almost featureless, as if still in the making, with an aspect of monotonous grimness. The edge of a colossal jungle, so dark-green as to be almost black, fringed with white surf, ran straight, like a ruled line, far, far away along a blue sea whose glitter was blurred by a creeping mist."[48] Conrad's book follows a voyage up the Congo River in the heart of Africa. What begins in adventure changes, throughout the novel, into something darker—a sense that the grim territory into which they are going is not so different from London; that imperialism has enacted a cost too great to be reckoned with. One of the novel's principal characters, the ivory trader Kurtz, dies whispering the words "the horror, the horror"—and by the end of the novel, that dismal pronouncement might be leveled at anyone or anything, from the travelers to the colonists, from the jungle to the city, or even (as is more likely) at the man himself, the dark heart of a dark heart. Mungo Park died in the country of Houssa after a local chief determined to keep the gifts Park brought to the king for himself. Reports suggest that Park leapt into the river with his companions. His body was never found. His son Thomas risked all to find some word or some remains, and he, too, vanished, never to return. Other attempts were tried, writes Verne, "all were destined to fail."[49] The Victorian theologian George MacDonald would write, "In whatever a man does without God, he must fail miserably—or succeed more miserably."[50] By 1914, nearly 90 percent of the African continent had been colonized and claimed by European countries, a conquest that destroyed nations and uprooted tribes, and that led, by politically frayed routes, to the First World War. George Shattuck Morison's prediction that "Savage and barbarous people disappear before the stronger arms of the more civilized"[51] rings both true and hollow. The enlightenment sciences did not conquer the Dark Continent, greater and more destructive weapons did, and it becomes impossible not to shudder at the thought of such dreadful "civilization."

The Victorian Age could no more see that threat than they could the advent of steam engines and automobiles. Darkness could be conquered. Would be conquered. Had been, even, and the bright light of a new industrial age scintillated with promise. The new age and its new scientists would prove a different breed, and they would face different enemies. Captains gave way to engineers, the eighteenth century's William Hershel to his son's revolutionary discoveries, and Humphry Davy to a man like Charles Babbage and a woman, Ada Lovelace, as they moved from mechanics and mathematics to the rust-hued dawn of computing. This new world is the world steampunk claims as its home territory—shiny gears, corsets, top hats, and engines. The Victorian Age bursts with Victorian industry. But the new world inherited a new kind of demon too. Just as radical skepticism and doubt arose from the same birth chamber as Newton's proofs of God, dread of privation, poverty, and loss claws its way forward in the very shadow of production. Heroes and villains, greed and debt, murders and medicine men, discoveries and detectives populate this history, and steampunk science (in fact and fiction) struggles to thwart old chaos in new guises. The Enlightenment was over. The Victorians had arrived.

And the game was afoot.

# PART THREE

"It is easy to understand that, when the new epoch is fully developed, all physical work may be dependent on inanimate power. It is easy to see that this means the concentration of enormous masses of power where power never could be had before; that it means the subdivision of power into units of a minuteness hard to conceive; that it means the unraveling of mysteries which have never been solved; that it means the construction of works of a magnitude before which the greatest monuments of antiquity become insignificant. The fighting ship of to-day is a floating machine-shop, though its crew of mechanics are confined as completely as the chained rowers of a Roman galley. [. . .] The camels of Persia will never again confront the elephants of India; fortifications will be factories filled with power. [. . .] It is interesting to speculate upon; it is foolish to prophesy about: these achievements are too close at hand for us to waste time in guessing what they will be."

—"The New Epoch and the University,"
George Shattuck Morison

## FIVE

# The Scientist and the Engineer

I had a dream, wrote Mary Shelley, "I saw the pale student of unhallowed arts kneeling beside the Thing he had put together. I saw the hideous phantasm of a man stretched out, and then, on the working of some powerful Engine, show signs of life and stir with an uneasy, half-vital motion." [1]

The mad scientist, with his mad devices, has become for us a signature of both science fiction and steampunk; wild looks, secret plans, and a genius turned toward dark ends. But science fiction frequently brings us *two types* of scientists, corresponding to two methods or "vectors" of science: those whose principal aim is discovery, and those whose principal aim is invention. "Discovery assumed reorientation of human knowledge," writes professor and co-editor of *Science Fiction Studies* Istvan Csicsery-Ronay Jr., while

"Invention is the active intervention."[2] We can trace the two lines of thought backward as well. Newton wanted knowledge for his own stock and store, to break through the mysteries of the universe and learn the language of God for himself; Leibniz instead invented a language for calculus that could be used and shared. Galvani wanted to learn the secret of animal electricity for its own sake; Volta wanted to build batteries to give people power wherever they needed it. The search after power might have begun in the scientist-philosopher's fascination with electricity, but its application—the only reason any of us have a use for it at all—has always resided in its *manufacture*. And so we've come back, at last, to where we began: the little book on the high shelf that proclaimed a new epoch on the coattails of that very principle.

James Watt developed his first steam engine in 1763, the first machine that could, by its design, manufacture more power. Humphry Davy might have used Volta's battery to power his arc light, but even he understood the limits of its utility: great stinking stacks of metal and acid could never power the future the way a steam engine could, and *utility* became the ultimate key to all the Victorian revolutions to come. The steam-powered looms replaced the need for weavers, turning cottage resources into massive industries—and turning a place like Manchester, England (or Newcastle, or Lancashire) into an industrial center. A century before, greatness meant increase of empire through exploration and coloniza-tion, but the nineteenth century opened a new portal to commercial and national success, and it rattled and clanked through the doors of invention. The world (and most particularly the paying revenues of political and national interests) had grown far less interested in what a discovery told man about himself—they wanted to know *what does it do?* Here and now, in the present mortal world, what can these things accomplish? "Seventy years ago," wrote George Shattuck Morison, speaking of the dawn of the Victorian Age, "Engineering was defined as the art of directing the Great Sources of Power in Nature for the use and convenience of man."[3] It was, he maintained, "limited only by the progress of science," and "its scope and utility would be increased with every discovery and its resources with every invention."[4] No one wanted a Newton, a man with knowledge locked up in his head. "The men who today are to direct the great powers of nature," Morison concludes, are the *engineers*—the very mechanical workmen of greasy toil that Descartes's contemporaries sneered at. And better, "the greatest engineer is not the man who knows the most, but the man who, when confronted with a new problem, can best grasp the novel subject, and whose judgment will most correctly approve or condemn his solution of it."[5] In other words, the new heroes weren't genius minds working away at solutions no one understood—they were everyday men and women with quick wits, oiled hands, and the ability to turn the abstract to practical use. So, even as the term "scientist" was coined by William Whewell in his 1833 review of Mary Somerville's *On the Connexion of the Physical Sciences*, a new breed of scientist was already on the rise. "The railroad has replaced the stage coach; the steamship has supplanted the graceful sailing vessel; and the telegraph has supplemented the laggardly mail," writes Morison, and "All this has been the work of the engineer."[6]

Morison lived his life in the roiling, rollicking industrial revolution of the latter Victorian era. He'd seen some of the greatest and most alarming shifts of any generation (excepting our current digital one): from horses to steam engines, from letters to telegraphs, from complete ignorance about infection to the establishment of germ theory. The greatest threats, in Morison's estimation, came from within, not without: "from the poisons, both moral and physical, which endanger concentrated populations," from bad air, bad water, and bacteria to bad construction and corrupt administration.[7] For Morison, anyone having to do with the steel bars and wood beams of the utilitarian and practical were by their nature engineers. He counts medicine, electricity, even the gas light of Humphry Davy or the achievements of bacteriology as but the further extension of the engineer—the extra fingers and toes of a vast mechanism that begins, in its way, to resemble the clockwork of *Whitechapel*. Humphry Davy's death signaled its beginning; his vacant chair at the Royal Society ushered in the first real struggle between old and new, between those who would discover and those who would invent, between the thing itself—power, bright, and terrible, electricity arcing and dazzling in its cosmic, godlike halo—and the thing that made it possible: the grit and grime of mines, the sludge that oiled the gears and regulators. In the rapidly industrializing future, time is annihilated and space contracted by trains and travel and telegraph wires, and the Victorians fought their demons from both sides: industry against poverty, waste against want. But the most interesting contest of all erupted over who controlled scientific knowledge, who had a right to participate, and who would get to direct the future (for better or worse). This is a story of engines. It's also a story of power. And we can learn almost as much from the machines that failed as from the ones that succeeded.

When we think of "engineering power" today, we might think of huge stacks and cylinders, spinning turbines, coal elevators and rails and earth movers the size of cities. Taichung Power Plant in Taiwan is the world's biggest power station—but England (motherland to most steampunk Victoriana) is home to its own megalith, *Drax*. Drax bristles with 139 conveyors, 12 cooling towers, 6 boilers, 200 railway wagons, and 1,800 miles of steel tubing. The manufacture of power on such a scale, to homes across a nation by virtue of connected wires, could scarcely have been imagined at the start of the Victorian age—but the seeds were planted, and germinating back in 1807, when Davy's arc lamp created a streak of light only four inches long, and took the power of two thousand voltaic cells to do it. By 1881, Paris hosted the first International Exposition of Electricity, featuring Zénobe Gramme's dynamo, Edison's light bulbs, Alexander Graham Bell's telephone, electric networks, electric tools, and electric trams. It would be followed by the electrically (and politically) charged World's Fair of Chicago in 1893. But the electric revolution began with far less fanfare, with the struggle of four leading minds over a singular problem, and with the greatest machine that was *never* built.

## The Little Engine that Couldn't

*"Have you never heard of Lady Ada Byron [Lovelace], then? The Prime Minister's daughter, and the very Queen of Engines! [. . .] Ada Byron, true friend and disciple of Babbage himself! Lord Charles Babbage, Father of the Difference Engine and the Newton of our Modern Age!"*

—William Gibson and Bruce Sterling,
*The Difference Engine*

I want to return for a moment to the Neverwas Haul described in the perambulation, part steamship, part automobile, all fantasy. The Track Banshees of the Nevada desert built the vessel to bring some of that fantasy into reality, to toy with the edges of space and time and to wonder about possibilities. *What does it mean to change history?* they ask—what does it take to build tomorrow's *unknown* from today's *known*, to blow upon that unsteady spark until it ignites? What-ifs light the steampunk world with diffusions as bright as Davy's arc lamp—but history had no Track Banshees. The Neverwas *never was,* and despite launching one of the best known examples of twentieth-century steampunk and igniting the imaginations of AI enthusiasts, the world's first calculating machine, the difference engine doesn't exist in history either. The designs weren't fully executed until 1991 at the Science Museum of London. The most interesting part of this story, however, isn't that the difference engine failed—it's that it *should have succeeded.*

Charles Babbage had a full head of steam. Described as a polymath, he had all the trappings of a young Newton: wealthy, elite, educated, and left (perhaps too much) to his own devices. His subject, like Newton's, was mathematics, but he arrived at Cambridge determined to upend the Newtonian system of calculus. Newton's complexity could not compete

---

\*    It might be better described as an OnlyNow. The physical machinery exists, but its historical context is fiction.

with continental mathematics, based as they were on the far more useful Leibniz calculus, and in 1812, Babbage established the Analytical Society to challenge it with his closest friend, John Herschel (son of the famous and eccentric astronomer, William). Babbage was bold, noisy, blustery, and rich enough to avoid most trouble. He joined a ghost society, he published tracts on religion dangerously close to blasphemy; he drank and gambled and otherwise led a life in keeping with the eighteenth-century genius. John Herschel, by contrast, was solid and sound—hard-working and brilliant in his own right, tutored generously by his father and guided by one of the century's most undervalued thinkers: Caroline Herschel, the Lady Astronomer. Despite their differences, the two were fast friends and eager for wider spheres than the stony and rule-bound Cambridge provided. In 1821, they traveled to the continent together, rogues on the trail of nature's hideouts, where they met Alexander von Humboldt, studied geology, and made excursions to the Alps (and the same glacier traversed by Victor Frankenstein's creature).[8] They returned home by November, and later that same year, the two of them stumbled into a problem without a solution—and the very beginning of a grand idea.

Stars don't always align. The story has it that Charles and John stayed late, poring over columns of numbers representing the position of stars in the sky at regular times through the year. Two different clerks compiled them faithfully, but the reports didn't match and calculation errors mounted. "I wish to God," complained Babbage, "these calculations could be done by a steam engine."[9] In June 1822, he presented a small model of just such a machine to the Royal Astronomical Society in London, an organization he founded, and which included as guest members John's aunt Caroline Herschel and the mathematician and astronomer Mary Somerville. His announcement mimics, in its breezy confidence, the style of Humphry Davy: "I have contrived methods by which type shall be set up by the machine in the order determined by the calculation. The arrangements are such that . . . there shall not exist the possibility of error in any printed copy of tables computed by the engine."[10] At its most basic, an electric calculator, it presaged much more. In 1825, Babbage's interest was piqued by electro-magnetism—and Herschel claimed that the strange phenomenon would lead the sciences of the new age (if only they could understand what it was). In its mythic properties, electromagnetism resembled Galvanic electricity, the concept so chased by Davy and the "Royal Society Hounds" in 1800.[11] At the time, Davy still held sway, though his powers were waning. In his service, however, was a young man named Michael Faraday, who had spent years laboring over Davy's experiments, traveling with him, putting up with his humors and ill treatment by Davy's wife. He was, by all accounts, a star on the rise. Despite this (or more probably because of it), Davy blocked Faraday's election to the Royal Society *eleven times*. A more disturbing story suggests that Davy engineered an experiment meant, literally, to blind the young Faraday by suggesting a lethal mixture of chemicals that exploded in his face once heated.[12] Holmes calls the contest one between "sorcerer and apprentice," but Faraday would climb well beyond Davy's shadow. In 1825, he took the position Davy held when he introduced the arc light: director of the Royal Institution. Modest, spiritually minded, monogamous to a fault, he doesn't fit the "genius" model of extravagance, but his lectures scintillated just like Davy's. Faraday took the stage as a shining light, described by onlookers as somehow unearthly, raptured: "his enthusiasm sometimes carried him to the point of ecstasy when he [. . .] lifted the veil from [Nature's] deep mysteries. His body then took motion from his mind; his hair streamed out from his head; his hands were full of nervous action; his light, lithe body seemed to quiver with its eager life"—and the audience, the onlooker proclaimed, "took fire with him, and every face was flushed."[13] The interaction between Michael Faraday and Charles Babbage, writes K. K. Schwarz, offers a key intersection for combining fields: science, industry, economics. But though the two of them are, respectively, inventors of the modern age—we only see it by detecting the current of history backward. Their lives may have been linked by "personal friendship, scientific interest, and a wide circle of friends," but they are rarely mentioned together, their separate works standing as evidence of minds, even genius minds, working apart rather than in collaboration.[14] But they do represent the gentleman of science at a pivotal moment, when the battle lines were being drawn between old and new.

Humphry Davy died in 1829, leaving the chair of the Royal Society empty and open to contest. Herschel and Babbage, both now of age and in the midst of their own scientific accomplishments, readied themselves for a fight—not with each other, but with an old establishment of old thinkers and old thinking. John, inheritor of his father's ambitions and his mother's wealth, and just as carefully raised on his aunt Caroline's principles (she was, after all, a discoverer of comets and the first female astronomer to be paid a living by the crown), seemed like the most natural choice. Faraday had refused to stand for the chair, preferring his position as director, and Babbage, member of the "Ghost Club," graduating without honors, and otherwise reckless, would not even be considered for the post. In the end, however, even Herschel would not make the cut—defeated 119 votes to 111, the young scientist lost to the Duke of Sussex, who actually knew nothing about science *at all*. He was, however, helpfully related to King George IV.[15] The nature of the loss led to open revolt among the young heads at the table, and some left the society altogether. Shortly after, Babbage published *Reflections on the Decline of Science in England*. Science, he argued, must be more than simple observation—it needed training in the skills of science, in self-control and dedication, and critical interpretation. Babbage's "manifesto" would become the foundation, there and then in 1831, of a new kind of unified society: the British Association for the Advancement of Science. Faraday didn't leap at the new model, though; instead, he encouraged Dutch chemist Gerard Möll to write a reproof of Babbage's bold assertions.[16] It would be Herschel who came to Babbage's aid, offering his own manifesto about the inductive method—about the testing of hypotheses, free scientific enquiry, and the investigation of the unknown.[17] It has a democratic flavor, and the British Association began to attract national attention by 1833. Even so, it would never fully overturn the Royal Society, and it did not encourage the government to increase its funding of scientific schemes, either—including

Babbage's difference engine. He abandoned the project 17,000 pounds in debt. The year was 1833, and Babbage was broke and bitter. In keeping with his usual style, his answer was to throw another of his intellectual soirees (for which he'd grown famous), and to invite the well connected, well educated, and well-to-do. One such guest was the mathematician Mary Somerville, and she'd brought with her a young pupil named Ada. It was a fortuitous meeting, because Babbage needed a new partner.

Ada Lovelace: only (legitimate) daughter of the mercurial poet Lord Byron, ostensibly responsible for the first written computer code, and a well-connected visionary of the mechanical future and in constant poor health. Tracking exactly who Ada was and what she achieved is like catching an oiled fish, not because nothing may be said of her but because everything might be. Both calculating and incalculable, Ada has turned up repeatedly in modern steampunk and science fiction, from Gibson and Sterling's *The Difference Engine* to Sydney Padua's recent graphic novel *The Thrilling Adventures of Lovelace and Babbage* [Fig. 12].* Christened Augusta Ada King-Noel, Ada lived estranged from her wayward father with her mother, the countess of Wentworth. Annabella Wentworth saw in Byron's poetics the rash insanity that drove him from her (they separated when Ada was only one month old) and determined that Ada's life would be very different. She would be nurtured on "true science," by which she meant the safety and utility of numbers, those same jots and dashes that described the heavens and ordered the world. The countess, when mentioned at all, receives a bare footnote in history as stiff and inflexible, but Annabella rose from a long tradition of women trained above their station, whose minds sought to build empires they could inhabit separate from bodies that could be mocked and traded upon by a world of men. It isn't surprising that she would have chosen Somerville as a tutor for her daughter, the foremost scientific woman of the age, and the same who was first to be admitted to Babbage's Royal Astronomical Society. Ada Lovelace clung to numbers not only as vocation,

---

\*     With thanks to the generosity of Sydney Padua, author and illustrator.

but also, as described by Iwan Rhys Morus, for physical salvation; she saw the future of machines as a future of body exploration—what might be learned in this "living laboratory" for the investigation, particularly, of electricity?[18] "It appears to me that the first thing is to go through a course of Mathematics," she explained by letter, including in her repertoire Euclid, arithmetic, algebra, and also Newton's treasured astronomy and optics. "I do not anticipate any serious difficulties; here I am, ready to be directed!"[19] She meets Babbage shortly after his plans to transform the Royal Society into a place for young ideas had all but collapsed, a talkative teenager, full of energy and a quick appreciation for the possibilities of the machinery.[20] Babbage dubs her his "Enchantress of Numbers," and by the early 1840s, they had become fast friends and correspondents, trading letters on the next phase of Babbage's calculating engine—and more importantly, how to sell the idea. It begins, like nearly all the other technological turning points, with a publication.

An Italian engineer, Luigi Menabrea, had published on Babbage's Analytical Engine. He had heard Babbage's lectures at Turin, and the work offered the first in-press description of the device (not delivered by Babbage himself). There were two problems with the publication, however; first, Menabrea was a virtual unknown at the time. Babbage had hoped the more esteemed Giovanni Plana would write the paper (and lend it his seal and credibility). Second, the paper was issued in Italian. If Babbage wanted to gain the attention of the world, particularly the English-speaking world, at least one of these disadvantages had to be surmounted—and Lovelace was more than capable of the task. In 1842, she published the translated memoir as *Sketch of the Analytical Engine Invented by Charles Babbage*, and with it, documentation that has been considered by some to be the most important contribution of the early computing age: an *algorithm*, the first software program ever written. Babbage had already acquired a series of punch cards, slips that could be fed into the new automatic Jacquard Loom for weaving. With some adjustments, the cards could be punched to program a very different sort of engine; as Ada puts it herself, "the Analytical Engine weaves algebraical patterns just as the Jacquard loom weaves flowers and leaves."[21] The punch cards operated like a more advanced version of the Writing Boy's replacement alphabet; as Babbage writes in a letter, "the system of cards which Jacard [sic] invented are the *means* by which we can communicate to a very ordinary loom orders to calculate *any* formula however complicated."[22] Lovelace's own breathless prose offers greater possibilities: "A new, a vast, and a powerful language is developed for the future use of analysis, in which to wield its truths so that these may become of more speedy and accurate practical application for the purposes of mankind than the means hitherto in our possession have rendered possible."[23] It wasn't a mere calculation machine they wanted; no! Why not follow what Leibniz himself had so long ago suggested? Why not build a machine to analyze and solve problems? Babbage's "Analytic Engine" would be the greatest feat of his day, a punch-card apparatus and mechanical storage on 50,000 cogs—what Holmes calls "the genuine equivalent of a modern computer's RAM memory." Lovelace saw the future the same way George Shattuck Morison did, but nearly fifty years before him. The future belonged to practical machines, and to the engineers who could build and run them. The machine, however, was not to be (not outside of fiction, anyway).

George Zarkadakis's 2016 *In Our Own Image: Savior or Destroyer? The History and Future of Artificial Intelligence* describes the way this singular achievement has fascinated historians and fiction writers alike. "How would the world be today," he asks, "if the British Treasury had not stopped funding Babbage's dreams and designs?"[24] Would he and Lovelace have ushered in the computing age a hundred years earlier? In the alternative history novel by William Gibson and Bruce Sterling, also called *The Difference Engine*, Babbage and Lovelace do indeed usher in a technological age nearly one hundred and fifty years before the first computer, ENIAC, was invented by J. Presper Eckert and John Mauchly in 1946. In this future, mass production of computers occurs by the mid-nineteenth century with the rise of information technology, while classical literature and humanities are abandoned in favor of engineering and accountancy—and Lovelace remains Ada *Byron*, the "queen of engines." But the answer to the question "what would happen if" Babbage had succeeded in getting additional funds for the difference engine is probably moot. Even with his personal capital to spare, he had not been able to make good on his plans, and moved from one idea to the next rapidly, abandoning the difference engine in favor of the analytic engine, then moving on to plan Difference Engine No. 2. Author and illustrator Sydney Padua describes him in an National Public Radio interview as a "blend of Mr. Pickwick [from Charles Dickens], Mr. Toad, Don Quixote, and Leonardo da Vinci."[25] He had a brilliant but erratic mind, one of the many reasons he'd not been considered for chair of the Royal Society and likely responsible for his inability to shift that body toward his aims and ideals. Lovelace is the hero of Padua's novel, a bright and expansive adventure that turns history's Augusta Ada King, Countess of Lovelace, into a lean, lithe, pipe-smoking, pants-wearing adventuress. The boisterous Babbage (who might have been as mercurial as Lovelace's own poet-father) is rendered a little less grand that he probably would have liked; it is the adventures of Lovelace and Babbage, and not the other way around, after all [Fig. 13].

The real-life Lovelace is hardly less fascinating. "The Devil's in it," she boasted after her successful translation, "if I have not sucked out some life-blood from the mysteries of this universe, in a way that no purely mortal lips or brains could do."[26] Babbage aimed for a calculating engine; Ada aimed much higher. The algorithms opened up possibilities that set her mind alight—and she willed and wanted almost entirely without direction, like Victor having discovered a secret but with no way to put it in practice. She wanted "cerebral phenomena such that I can put them into mathematical equations; in short a *law* or *laws*, for the mutual actions of the molecules of *brains*," that were "equivalent to the *law of gravitation* for the *planetary* and *sideral* world."[27] She felt certain there were connections that somehow bridged the gap between material and immaterial, among intellectual, moral, and religious understanding and the electrical impulse of body tissues. The physical universe and the teeming millions of thoughts in the teeming millions of brains must all be one; we need only map them as we had the great dark continents. Had she been alive to know Kepler, she might have found a kindred spirit, but she knew Babbage instead—and Babbage would lay the foundation of computing largely without her. William Aspray, author of *Computing Before Computers*, suggests that Ada never really programmed anything at all, while James Essinger's work claims that she launched the digital age. In her later years and in increasingly poor health, her interests in electricity would become profoundly personal, linked to a wave of (usually unwholesome and often dangerous) medical-electrical treatments rather than to building computing machines—something we'll return to in later chapters. Could she not build a better body—or even a mind? It returns us to Hiroshi Ishiguro's aims to replicate himself in robotics, and to the idea that even inanimate things may have souls. Did the real Ada Lovelace "code" or not? It remains difficult to say, but no one can argue about her *influence*. "Can I do that, truly?" asks Sybil, the protagonist of *Difference Engine*, about a woman's right to be a *clacker*. "Can a girl do that?"[28] Ada's powerful connections gave Babbage something he hadn't been able to pull together entirely on his own: publicity, finance, and *noise*. Ada was captain to Babbage's engineer, the salesman with a taste for the carnivalesque and the pedigree to rub shoulders with the scientific elite. And no engineer ever got very far without one.

Why, then, doesn't the difference engine succeed? Scientific progress in electricity continued; in 1833 Faraday discovered electromagnetic induction, or the production of force across an electrical conductor due to its interaction with a magnetic field; engines and hydraulics and telegraphs would be developed in the ensuing years. But of the analytic and difference engines: nothing. Babbage understood the punch cards; he'd solved the first problems of the machine and even built small models that worked. Menabrea grasped the concept that mathematicians might "execute, by means of machinery, the mechanical branch of these labours, reserving for pure intellect that which depends on the reasoning faculties"—that the hard and arduous business of calculation might be gotten at by engines with less time and trouble. Ada's translation provided the tables of numbers, explained how the numbers could be worked into the machine by using theorems and laws on punch-card programs, and her further speculations explain how an analytical engine would operate. Between the two of them, they had all the necessary connections to intellectual and high society, including the support of John Herschel and (if not enthusiastically) of Faraday. They conversed with other great minds at the British Association meetings, which gained in importance and influence, and they promoted all their doings through publication and display. Gibson and Sterling—and Sydney Padua—begin their alternative narratives in a world where the success of the engine is taken as first principle. But what makes a success? And why, with so much on their side, did Lovelace and Babbage ultimately fail? To answer the question, we have to start down a different path entirely. Not surprisingly, it has everything to do with purpose—and with dread.

## The Tale of a Ship

In 1868, the United States Navy commissioned a ship called the *Wampanoag*. The ship still retained the mast and sails, but unlike all other vessels in the fleet, she was propelled by *steam*. An engine with two cylinders 100 inches in diameter punched out power with a four-foot stroke that turned giant wooden gear wheels and a propeller 19 feet wide. A 60-pound gun, two 100-pound guns, ten 8-inch guns, and four howitzers completed the armory, and she could compete in heavy seas at 17 knots—the fastest in the world.[29] She'd been built by an inventor named Benjamin Franklin Isherwood, and she accomplished things that most engineers thought impossible. She was too long, the ratios were off when compared to other ships, her shape would cause her to roll over in the water, she would be hard to maneuver—or so claimed the board of naval officers sent to examine the *Wampanoag*. They listed out six points in their report, called the ship a miserable failure, and had her decommissioned . . . *even though none of their predictions came true.* In fact, the ship succeeded in every possible way. She beat all the other navy ships at their own game: swift, steady, efficient, and easy to steer.[30] But though the officers stood on her decks and saw her in action, they still recommended she be scrapped as a failure, posthaste. Here we had a ship, far more excellent than any of her day, and she never made official maneuvers in the capacity she'd been designed for. Like Babbage's difference engine, all the means of success appeared to be on her side, and on the side of her inventor, Isherwood. And like the difference engine, the *Wampanoag* never catches on.

The *Wampanoag* story appears in a lecture by Elting Morison, nephew of George Shattuck Morison. "Now it must be obvious that the members of this Naval Board were stupid," he writes. The men who opposed the future did so on bad principles, because they feared the future or had fixed attitudes, he supposed; they might be compared to the intractable Royal Society and the refusal of funds for new inventions by new minds. Steam engines disrupted the world they knew, the world of sails and masts and canvas and rigging. Steam engines needed boilers, dark hovels, and grime-faced engineers to run them. Resistance to change is built into our very genes. But, then again, so is the desire for the new. What is it that determines whether a new thing will be embraced or tossed to the side? "It was at this point in my research that I began to be aware of a growing sense of dis-ease," Elting admits. The officers had described the ship the way Frankenstein's monster described himself: she was "too much of an abortion"—by which they meant an anomaly, anathema, a thing that did not fit and should not be.[31] The Civil War, you see, had ended. And, in being a ship like no other ship, she had no opponent and no purpose. That is, as a ship of war in a time of peace, the *Wampanoag* had nothing to *do*. Elting ultimately determines that the "stupid" officers are right, in a way. The new ship was too new; she'd arrived too soon and was "a destructive energy" in their society. To return to the very beginning of this book, a machine, if left to itself, will "establish its own conditions," and bend the man to the machinery rather than the other way around.[32] And that is fearful business.

In Gibson and Sterling's novel, the "winner" isn't Sybil or Ada Lovelace/ Byron, but the tech itself. It ends ominously enough, in stilted, stuttering, stream of consciousness. "It is 1991," the narrative voice explains, "it is London. Ten thousand towers, the cyclonic hum of a trillion twisting gears, all air gone earthquake-dark in a mist of oil, in the frictioned heat of inter-meshing wheels. Black seamless pavements, uncounted tributary rivulets for the frantic travels of the punched-out lace of data, the ghosts of history loosed in this hot shining necropolis." From the disconnected images, emerges "a dry foam of data, its constituent bits and motes [. . .] swift tireless spindles flinging off invisible loops in their millions." Data and networks, communication hubs and wireless signals: even in the alternative history, a structure must undergird computational logic. Lovelace might be able to *explain* the difference engine and analytic engine: one calculates arithmetic and is finite, while the other has "no finite line of demarcation which limits the powers of the Analytical Engine."[33] Her work might negotiate the mathematics of program language, even the necessities of how to put data to use in the machine. Like Isherwood and his ship, Lovelace could easily explain what the engine could *do*—but not what it was *for*. The world did not need the difference engine. Not yet. What it needed, what the rapid industrialization of Britain required, and what leading figures of the British Association most *wanted*, were engines of an entirely different sort.

In 1836, the British Association for the Advancement of Science upheld a man named Andrew Crosse as the perfect picture of the scientist. He was "disinterested and humble," solitary, quiet, and in search of "nothing beyond nature's truth."[34] He spent his time coaxing crystals by passing electrical current through various substrates, and strung his property with wires for collecting atmospheric electricity. Fifty Leiden jars packed the music room "flashing and crackling [. . .] with playful intermittence all the livelong night."[35] But this vision of science—divorced from the material, depoliticized, and shut away in the private laboratories of private men—was about to be electrified out of existence. In 1837, Charles Wheatstone and William Fothergill Cooke patented the electromagnetic telegraph, building on work that Faraday had begun but for the first time *applying* it. Long-distance communication became the first Victorian electrical industry, and launched an age of entrepreneurs[36]—and in 1838, the year of Queen Victoria's coronation, the British Association met in Newcastle, a town on the rise and soon to be a center of industry. Attending the meeting were the scientific greats, including Babbage, Herschel, and Whewell, as well as a singular man of singular (pre)occupation: an engineer, armament magnate, and inventor, he would (with the near-mythic Isambard Kingdom Brunel) build the machines that worked—the engines that *could*.

## Newcastle's Engineering Giant

The steampunk aesthetic—the blowing whistle, the whirring engines, the clockwork cams, and retro-futurist Zeppelins with their pilots in frock coats and top hats—offers a whole world bright with color and alive with moving parts. But this extraordinary ethos didn't emerge from a vacuum (unless you count the vacuum tubes of the static electricity generator). Up to the 1830s, Britain's scientific endeavors rolled along in service of King George—rational, methodical, and if Faraday had anything to do with it, entirely and immanently respectable. There were showmen; there had to be. But there must also be more than plans; the show needed substance. It needed a man with invention on the brain, but with the ability to turn it to civic good; someone who charmed with his good nature, his intellect, and his eclectic habits; someone who would turn ideas into money and then back into more ideas. It needed a tall, elegant, top-hatted figure stepping from a train platform designed especially for the purpose of connecting his estates to the rest of the wide world—it needed, in other words, Baron William George Armstrong.*

---

\* For a comprehensive biography of the "magician," see Henrietta Heald's *William Armstrong, Magician of the North.*

Conceivably the only reason Armstrong has never been rendered as a fictional hero of his own steampunk story is because we already have Verne's Captain Nemo. Armstrong doesn't captain an electric submarine, but he does build the first electric house complete with lightbulbs (arc, and then incandescent), running water, central heating, dishwasher, vacuum cleaner, passenger lift, fire alarm, electric dinner gong, and self-turning kitchen spits. As obsessed with clockwork as Phileas Fogg from Verne's *Around the World in Eighty Days*, he observed exact (and exacting) timetables on his guests. "There are few deviations," wrote one guest, "Breakfast is on the table at 8 to a second! Luncheon at ½ past 1, tea at 5, and dinner at from ½ past 7 to ¼ past 8."[37] He had a reputation as a "modern magician," changing not only the technology but reforming the landscape; he dammed the Debron Burn waterway and installed a water-powered dynamo, supplying electrical wonders. "Ten thousand small glass lamps were hung from the rocky hillsides or upon the lines of railing which guard the walks, and an almost equal number of Chinese lanterns were swung across leafy glades" wrote contemporary visitors, "little lamps" all "hung like fireflies."[38] Armstrong had his own observatory for stargazing, but he attracted stars of another sort. Though descended from common stock, the son of a lawyer and grandson of a shoemaker, the engineer played host to the Prince of Wales and other dignitaries who longed to see the magic of Cragside house. To commemorate the occasion, the *Times* praised Armstrong as "a name which embodies all the great industries and interests of the city."[39] The prince paid the compliment, too, in his dedication of Armstrong park, twenty acres along the Jesmond Dene: "his name is known in the British dominions as that of a great man and a great inventor, but may safely say it is known all over the world."[40] He was likewise known for the Elswick Works, a huge and sprawling plant for the production of hydraulics, shipbuilding—and later, armaments. As Armstrong's premier biographer, Henrietta Heald, explains, the great engineer and "magician of the north" was known as equally for wringing "the blood and sweat of thousands of men [. . .] whose demands for improved working conditions he cavalierly dismissed" and for the creation, manufacture, and sale of the "most ferocious killing machines the world had ever seen."[41] Armstrong operates at the edge of hero and villain, remembered fondly as benefactor and friend, antagonized as a merchant of death. But most importantly, he rises as the first of those engineers George Shattuck Morison described as champions of a New World.

Civil engineering, says Morison, embraces all other types—and to build the very means by which engineering itself is carried out. In his swift and breathless prose, he lays out the groundwork for this feted designer of the future: "The civil engineer is briefly a man who, with knowledge of the forces and materials around him, uses that knowledge in the design and construction of engineering works"—his business is "to design the tools" of power, by power, so that power might be directed by man.[42] Even in Armstrong's lifetime, those who knew him claimed that he "cast his thoughts into iron."[43] If Babbage and Herschel longed for a day when young minds and active bodies would overturn a crusty Royal Society, here they had their wish—but though Armstrong would be made a member of the Royal Society, he was not, nor would he have considered himself, primarily a *scientist*. Granted, the term was only coined in 1833, a few years before the British Association meeting that saw all those upon whom Whewell might bestow it in the same place at the same time. Even so, before the 1840s, Armstrong was a lawyer secreting a passion for mechanics, engineering, and electricity. Sent to London's Temple district to learn law, Armstrong spent his Friday evenings at the Royal Institution, listening to the same dashing and charismatic Faraday lectures than inspired Ada Lovelace. Faraday's compelling lectures now included something more tangible; his discovery of electromagnetic rotation led not only to his invention of the first transformer, but also of the first *dynamo*, an electrical generator that used a commutator to produce direct current (DC). As DC, it didn't produce enough energy to be of much use, but it proved the first practical step for a mind geared toward electricity's possibilities. Armstrong made his own electrical discovery by 1840; in a letter to Faraday, he tells of an accident where a steam-engine operator received a violent shock from the steam itself.[44] Through a series of experiments, Armstrong isolated the phenomena: the steam electrified as it entered the atmosphere due to the action of friction as it emerged from the pipe. That meant *steam power* could be *electric power*—and he set about building the boiler that would change his career, the future of Newcastle, and the course of two wars. The manufacture of power had begun.

One dark night near Christmas, a deep frost set in. The dark streets ghosted white, breath feathered from chapped lips, and bodies huddled into winter wool. A young man named John Wigham Richardson waited in the dim audience of the Literary and Philosophical Society for a wondrous glimpse of the electrical machine: "It was a weird scene," he later wrote, "the sparks and flashes of electricity from the machine were, I should say, from four to five feet long and the figure of Armstrong in a frock coat [. . .] looked almost demoniacal."[45] We hear, in the description, echoes of Shelley's mad Victor; we see, in the flying sparks, a hint of wizardry to come. But Armstrong did not launch his career in thunder and lightning. He began (while still diligently pursing life as a solicitor) by studying *hydraulics*. We talked of Newton's year of magical thinking; Armstrong had a magic year, too; his greatest discoveries did not come then, but the tracks were laid. Victoria had taken the throne, and Armstrong had taken to water, specifically hydraulics. In 1838, he published an article about a water wheel in *Mechanics' Magazine*, suggesting that water power could be concentrated during descent to propel machines; the following year, a working model was installed at Watson's Works. Armstrong's lifelong friend and fellow engineer Thomas Sopwith called it marvelous in its simplicity, with water pressurized and applied to a piston.[46] He even speculated on its many uses—but, as with Babbage's difference engine, the hydraulic engine generated no real interest. Not then. He'd invented a hydraulic crane as well, able to lift weight with the action of a single piston. Again, no interest. In an early history of Elswick (1909), Alfred Cochrane remarked that the most "curious" thing about the cranes was that no one seemed to care for them. Armstrong invented, modeled, and then built devices whose utility we take for granted now as much as we do Babbage's calculating engine, but he suffered from the same trouble: the world didn't need to know what the machines did. They needed to know what they were *for*. Armstrong needed a public, and in December 1842, shortly after demonstrating his electric steam engine to the delight and terror of young John Wigham Richardson, the device appeared on display at the Polytechnic Institution. The feat had been accomplished by Thomas Sopwith, and by threats of taking the crowd-pleasing, spark-producing dynamo elsewhere. Like Matthew Boulton to James Watt, and with equal success, Sopwith gave Armstrong room to work wonders. A year later, the Lit and Phil Society lectures were so crowded that Armstrong himself had to climb in through a window to the stage—he appeared in print, jocularly touted a wonder-worker by the press, and awed his audience like a latter-day Humphry Davy, lighting a cannon with a spark from his finger to conclude the show.

The flash and bang may have been the chief attraction (and no one can argue the results were more interesting than a list of mathematical figures calculated by steam). But the difference between Babbage, Lovelace, and Armstrong goes deeper. Part of Armstrong's allure, writes Heald, was his ability to *explain*. Unlike the unspooling mathematical code annotated and translated by Lovelace, or Babbage's own bitter invectives against those who did not clearly grasp his work, Armstrong—a layman himself—spoke in laymen's terms. "Although not an outstanding orator," Heald explains, and without the stage charisma of either Faraday or Davy, "he had a clear, precise, authoritative method of delivery, shot through with wit and humour."[47] And he always had a ready audience. By 1846, at the young age of thirty-five, Armstrong was made a Royal Society fellow. Sponsored by Faraday and Charles Wheatstone (inventor of the telegraph), "What is the Royal Society for," they asked, "if not for such men as Armstrong?"[48] He was, by any measure, a remarkable man—he had a remarkable network, and a remarkable partner in his wife Meggie, who aided him in his projects and experiments. (Heald recounts a story about wood slats and lime that suggests the *Thrilling Adventures of William and Meggie* would be worth reading.) But he continued to work as a lawyer with his friend and partner Armorer

Donkin till 1845, until an unexpected set of circumstances, coupled with his head for business, created a true public need.

In this case, the need was water. Clean water. Charles Dickens's *Hard Times* provides one of the best fictional descriptions of the foul waters brooding about industrial centers, but the truth was often darker and harder still. In Mayhew's *London Labour and the London Poor*, poor river boys pick coal and debris out of the Thames, picking it out of the muck.[49] In 1858, the Thames became known as "the Great Stink," industrial waste and sewage combining with a singularly hot summer to create an oppressive and relentless stench that literally shut down the city of London. Michael Faraday staunchly supported water reform as early as 1855 and published a paper called "Observations on the Filth of the Thames" where he described "the whole river [as] an opaque pale brown fluid."[50] But the Rivers Pollution Commission of 1874 would sum up their Thames and Lea River findings as follows: "that the river receives the sewage from a large number of towns [. . .] the washings of a large cultivated land, and the filthy discharge from many industrial processes and manufactures"—and that it floated with excrement and animal carcasses.[51] Problems with the water system had become endemic, growing up in slime and sludge from the industries that fed national capital and growth. The stench alone, rot and filth and excrement combining with the waste of stagnation, would move legislation—but the smell was hardly the only problem, and London wasn't the only locus.

Cholera. Dysentery. Typhoid. For years, doctors had been helpless in the face of mystery illness. Fevers seemed to come out of nowhere, wretched stomach complaints, and worse. But the king of these epidemics, cholera, could kill in days. Cholera rendered the body viscerally incapable of taking in fluid; vomiting, diarrhea, severe dehydration—the hands and feet would wrinkle and shrivel, the skin would turn blue or green. Death occurred in the midst of twitching, cramping muscles, a vile, unsanitary end that spread through whole communities and felled seemingly healthy people overnight. John Snow, a physician in London, finally traced the source to contaminated water in 1855, but long before his discovery, residents and officials recognized the relationship between unsanitary conditions and disease. Where towns were growing quickly—towns like Tyneside and Newcastle—overcrowding and waste rapidly turned good drinking water into dangerous cesspools. Edwin Chadwick published a report titled *Sanitary Conditions of the Labouring Population of Great Britain* in 1843, and shortly following, William Armstrong (with Thomas Sopwith and building Richard Grainger) submitted his first proposal for a solution: he would convey clean water from Whittle Burn (a clean-running tributary west of Newcastle) to the residents of Tyneside. The project was an ambitious one: they must build a series of reservoirs and lay pipe through large tracts of land. But by 1848, the Whittle Dene Water Company laid more than twelve miles of iron pipe, two feet in diameter—the largest in the world.[52] The success certainly served the public good, but as Heald remarks, the continuous supply of water served something else too; Armstrong now had the means to make a case for hydraulic power. A lengthy lecture and a brilliant display of the hydraulic crane later, and W. G. Armstrong & Company began building the Elswick Works. Armstrong would not go back to law. He would become, instead, one of the foremost engineers of the Victorian Age by explaining not only what water power could do—but by showing, without doubt, what you could do *with* it. Technology thrills best when used against the disintegration we fear, the chaos, the dread. When we, that is, control *it* and not the other way around.

The Elswick Works became the single largest industrial plant of Great Britain, employing thousands. They constructed the first hydraulic cranes in 1847 for the Edinburgh and Northern Railway, then more for the Allenhead Lead Mines.[53] A few years later orders came from everywhere, most notably from a man who, even more than Armstrong, embodies the Victorian engineer: Isambard Kingdom Brunel. The man behind the Great Western Railway, the network of tunnels, bridges, viaducts, and all that linked the land between, dark double lines of rail from London to Bristol—Brunel offered something of the panache of Babbage in caricature. His most famous photo, one taken by Robert Howlett, features Brunel in disheveled gentleman's apparel and top hat: jaunty, determined, hands in pockets and cigar in teeth, Brunel stands astride the giant chains *not* of a railway bridge, but of a *ship*: the *Great Eastern*. And Armstrong's Elswick would get into the shipbuilding business too. And yet, neither the cranes nor the ships would ultimately make Armstrong's career. Instead, by the 1850s, he and Brunel would be designing engines of war, building big guns for the Crimean War and the siege of Sebastopol.

## A Natural Resistance

We like to think of flashes of genius, of thunderbolts like those described by Victor Frankenstein when he suddenly knew the secret of life, when the curtains were thrown back to reveal the great shining truth beyond truth. Or, if not genius, then the madness-induced vision of the castaway in *Moby-Dick*, "carried down alive to wondrous depths, where strange shapes of the unwarped primal world glided to and fro before his passive eyes; and the miser-merman, Wisdom, revealed his hoarded heaps [. . .] He saw God's foot upon the treadle of the loom."[54] But history's greatest inventions do not leap ahead of the rails. They do not, because they cannot—or rather, because we cannot. We can't want a thing we haven't a need for; the smart phone could never have come before the Internet. When Brunel built the *Great Eastern*, the 22,400-ton behemoth upon whose chains he stands puffing in his photo, it was the largest ship in the world. It also took three months to get her afloat, and by the end, she was more than bankrupt, her original company gone under the surface and the new owners at a loss about what to do with a great empty hull that no one wanted to ride upon. It might have suffered the same fate as the *Wampanoag*, except that in this case, there was now a need. Wheatstone patented the telegraph, and soon telegraph had become the most fundamental communication network of any kind—it revolutionized the world, and from 1866, it would connect the world too. Thousands of miles of transatlantic cable needed to be run, and for that, they needed the world's biggest steamship. Brunel has been lionized as a genius, a madman, a superhuman inventor—but he was, more than anything, an engineer building upon the engineering of others: ships to bigger ships, rails to the Great Western Railway, and rifles to the world's largest guns. Our human resistance is overcome only at cost, only by the slow building of engine upon engine; and after all, even the "clacks" of *Whitechapel Gods* creeps, rather than collides, a slow shifting of man into metal like the steady accumulation of Elswick Works on the river Tyne.

"The substitution of inanimate power for human labour," writes William Armstrong, "must unfortunately always be attended in the first instance with the evil of depriving individuals of employment."[55] And yet, "the general welfare [. . .] is unquestionably promoted in every instance in which we succeed in coercing insensible agents into our service." Man, says the magnate, "was designed to work by his head rather than by the mere strength of his arm" as he "continued to extend his dominion over the powers of nature."[56] It is right to suspect that George Shattuck Morison knew of these words—or else we are left to assume that engineers of his kind and temperament are cut of such similar cloth that even their hopes for the future use the same words and strike the same notes. Armstrong succeeds because he builds machines that the world already knows it wants: the drilling, punching, boring, tunneling, clipping, clanging, pressing machines. The machines that built cities and dug tunnels and laid rail. The machines that protected borders and the machines that crashed through them. When London's Tower Bridge opened in 1894, it was powered by Armstrong's hydraulic accumulators.[57] When the navy built warships in the second half of the century, they used the same technology. Step by step, the engineers laid a path that other engineers would follow; the steps themselves might be matters of minutiae. They may seem but the subtle shifting of gears. But what was true of Hauksbee's electric machine is just as true of those makers who came after: every new machine had its beginning in the last machine's ending. When the troops failed and flagged at Sebastopol, stymied by war technology that hadn't been updated since the days of Napoleon, both Armstrong and Brunel put their engineering facilities to the task. James Rendel, a civil engineer, worked over details for new artillery in an evening with Armstrong, blotting paper and sketches between them: "*You* are the man to do it," he exclaimed. And in June 1887, the *Illustrated London News* showcased Armstrong's 111-ton gun, crafted at Elswick Works. It shot 1,800-pound rounds at the rate of 2,020 feet per second; it could penetrate wrought iron to the depth of 30 inches from a thousand feet away . . . and most shocking of all in an age of close-range water warfare, it had an effective range of *eight miles*.[58] The gun served as the main attraction at the Royal Mining, Engineering and Industrial Exhibition in 1887, held in Newcastle to celebrate the Golden Jubilee of Queen Victoria's ascension to the throne—effectively also the anniversary of that fateful British Association Meeting at which Babbage, Armstrong, and others first gathered fifty years earlier. He had become the master engineer, a "magician" of electricity, and also the mastermind behind the deadliest weapon of its time. He would go on to supply armaments not only to the British government, but abroad as well—in fact, Armstrong outfitted *both sides* of the American Civil War. And he made a fortune doing it.

History always reveals patterns of resistance and change. Sydney Padua describes Babbage's engine as a "beautiful machine," almost an art installation. "He wasn't aiming at the practical," she explains; a device that—like Newton's language of God—meant to tell a man about himself. The difference engine was for thinking, not for using, and Babbage "withheld the machine because he didn't feel the world deserved his engine."[59] Why make for *them* what they could not understand? His language was math, and like Newton, he didn't think math belonged to everyone. It wasn't only the wider world's failure to see the utility of Babbage's machine—he couldn't see it himself. And this is nowhere more apparent than in his correspondence with Brunel, who was his longtime friend. In July of 1857, Brunel offered to help him get the difference engine into manufacture,

FIG. 1 (ABOVE): Hans Holbein the Younger, *Jean de Dinteville and Georges de Selve* ("The Ambassadors") © *The National Gallery, London.* FIG. 2 (BELOW): Athanasius Kircher published a similarly cosmic work, *Ars Magna Lucis et Umbrae. Dittrick Museum of Medical History.*

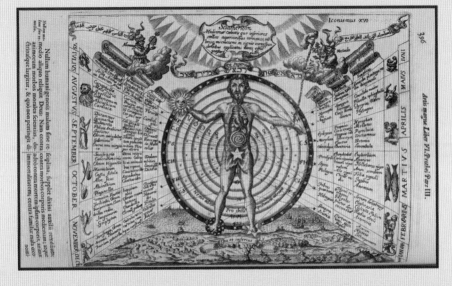

FIG. 3: Jacquet-Droz, *The Writer. Musée d'art et d'histoire, Neuchâte* *(Switzerland). With thanks to Dr. Christian Hörack, Conservateur d: département des arts appliqués and the generosity of the museum.*

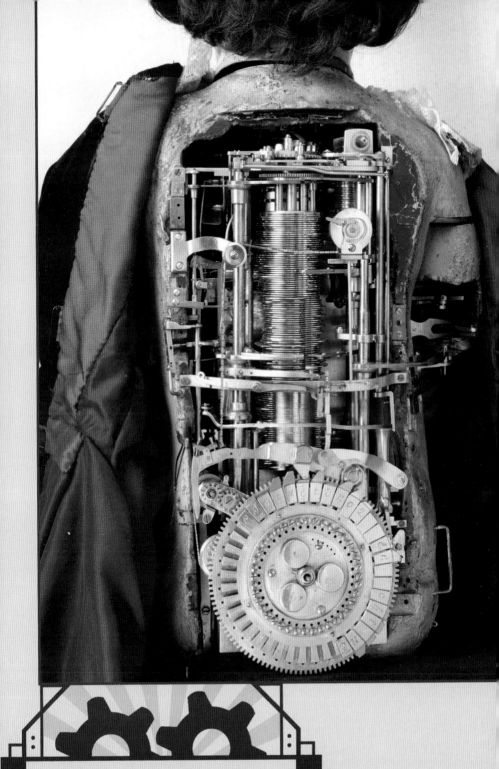

FIG. 4: Jacquet-Droz, *The Writer*, interior. *Musée d'art et d'histoire, Neuchâtel (Switzerland)*

FIG. 5 (BELOW): Birth mannequin of "phantom." *Dittrick Museum of Medical History. With thanks to Laura Travis, photographer.*
FIG. 6 (RIGHT): The electrostatic machine, *Francis Hanksbee. Wellcome Images, Wellcome Library, London no. M0012622.*

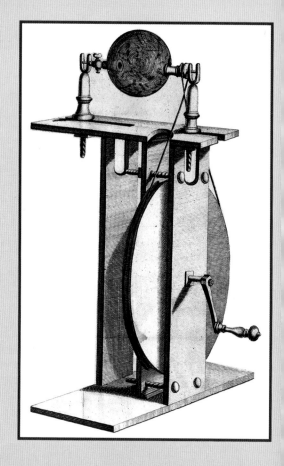

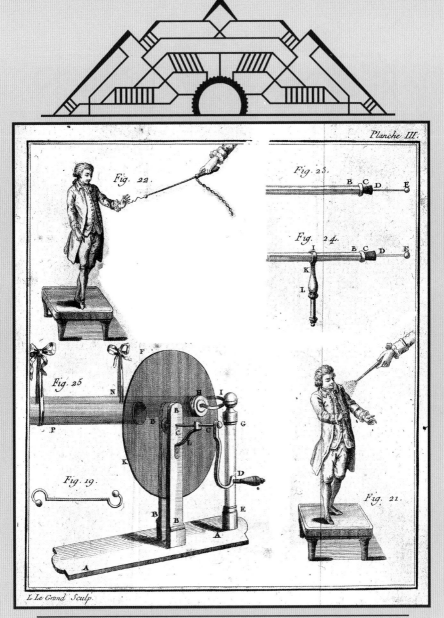

FIG. 8: (OPPOSITE): A paper figure, which would "dance" when charged with static electricity. With thanks to the Bakken Museum, Minneapolis, Minnesota. *Photography by Brandy Schillace.* FIG. 9: (BELOW): Illustrations from Fowler's book, demonstrating electrical galvanism on dissected frogs.

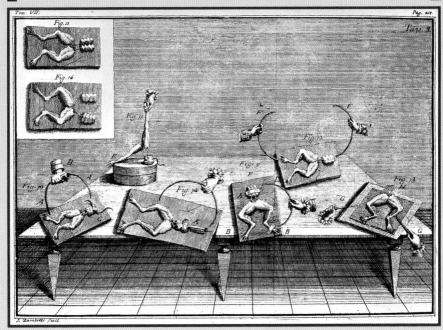

Fig. 9 illustrations taken from *Experiments and Observations relative to the influence lately discovered by Calvani and commonly called Animal Electricity* by Richard Fowler.

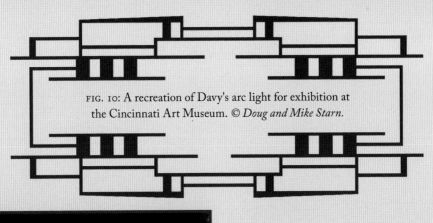

FIG. 10: A recreation of Davy's arc light for exhibition at the Cincinnati Art Museum. © *Doug and Mike Starn.*

FIG. 11: Frederic Edwin Church, *The Icebergs*, 1861, Oil on canvas © *Dallas Museum of Art*.

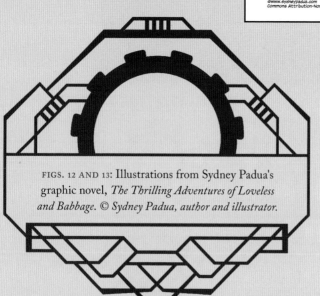

FIGS. 12 AND 13: Illustrations from Sydney Padua's graphic novel, *The Thrilling Adventures of Loveless and Babbage.* © Sydney Padua, author and illustrator.

FIG. 14: Nikola Tesla, with his equipment for producing high-frequency alternating currents. Inscribed "To my illustrious friend Sir William Crookes of whom I always think and whose kind letters I never answer! Nikola Tesla June 17, 1901." *Wellcome Library, London.*

FIG. 16: (ABOVE): A carbolic acid sprayer from the collection of the Dittrick Medical History Center and Museum. *Photograph © Laura Travis.* FIG. 17 (BELOW): Advertisement for an electropathic belt. *Photograph © Laura Travis.*

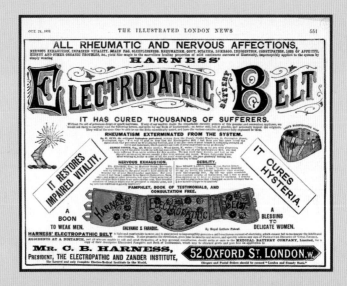

saying that once a workable system could be offered, "fresh wants would arise."[60] Build it, he suggests, and they will come. Except that they did not. And in 1914, when Lord Moulton visited an aging Babbage to see the finished machine, it still remained largely in Babbage's head: "There lay scattered bit of mechanism, but I saw no trace of any working machine. Very cautiously I approached the subject, and received the dreaded answer: 'It is not constructed yet, but I am working at it, and it will take less time to construct it altogether then it could have to complete [. . .] the earlier version from the stage in which I left it.'"[61] We can see in the machine's design the kernel of all the computers that run our world, but that leap wasn't made by Babbage at all.

At the dawn of the Victorian Age, Britain sought to build its industrial empire and also to protect it; the difference engine made no difference at all to those aims, but cranes, and railway tunnels, ships, bridges, and guns *did*. For all our latter-day steampunk interest in the disputed birth of AI, possibly the image that most captures the aesthetic of the real Victorian steam-driven imagination is that of Brunel before his shipyard chains, or of Armstrong demonstrating the steam-driven spark of electricity. The latter half of the Victorian Age would turn on the possibilities of inanimate power, and Armstrong predicted it would be *electrical power*. The biggest challenge would be to make it cheap and abundant. "We have firstly the direct heating power of the sun's rays," Armstrong explains to the 1881 meeting of the British Association at York, "but have not yet succeeded in applying [it] to motive purposes."[62] Instead, he insisted, we must work toward water, wind, tidal, and the "preserved sunbeams" of coal. What they needed to harness the black rock into power for everyone was means of conducting the electricity produced and a means of conveying it over longer distances than seemed possible at the time. Someday would come, he prophesied, when conveyance and batteries would change the world, when solar power might be possible, when waterfalls might be tamed, and when "an electric horse far more amenable to discipline than the living one may be added to the bounteous gifts which science has bestowed."[63]

Science, he says. Only we know already that this is not Babbage's science, and not Newton's either. The new masters of power would become prototypes of the steampunk hero, and be immortalized, too, in fiction as well as fact. And it's important to note that Armstrong's dream of water power would be realized, after all, only fourteen years after his address by a man as nearly mythic in status as the fictional heroes of Jules Verne: Nikola Tesla.

## SIX

## He Who Powers the Future

homas Alva Edison was high profile, high rolling, full of good humor and, not infrequently, practical jokes at the expense of others. He ate pie as a regular meal, barked orders, solved problems, and made brazen claims. In the retrospective of history, Edison was larger than life. Arthur Palmer, one of Edison's employees, said this at his passing: "He led no armies into battle, he conquered no countries, and he enslaved no peoples . . . Nonetheless, he exerted a degree of power the magnitude of which no warrior ever dreamed. His name still commands a respect as sweeping in scope and as world-wide as that of any other mortal."[1] Edison was born in 1847, less than a decade after Queen Victoria's coronation and an ocean away from Armstrong and Brunel. He held no advanced degrees and possessed no refined manners, but rubbed elbows with high society anyway. To his credit, he held over one thousand US patents; he invented the phonograph, the motion picture camera, and most famously (by hook and crook in a race against others), the first practical and mass-produced light bulb. Edison's methods were far from the finesse of earlier scientists and engineers, but he was shrewd and knew how to produce results en masse. In fact, he is credited with the dawn of the industrial research laboratory, a space where all men worked more hours than they slept, slaving away at patents that would never bear their names. Edison shines as a hero of George Shattuck Morison's engineering dream: a poor boy who pulled himself up by bootstraps and became the greatest mind of his generation. They called him the Wizard of Menlo Park.

The United States clawed its way out of civil war to face a rapidly industrializing world—and a rapidly expanding one too. By 1878, immigrants poured into New York Harbor, looking for work in factories and filling the tenements to the teeming millions. Like the underbelly of London, cities like New York, Chicago, and Cleveland boasted slums reeking of filth and rife with cholera and typhoid fever. But, as Jill Jonnes writes in *Empires of Light*, wealth poured into the city, too, and the "magnificent buildings of Wall Street" reflected the nation's growing economic power.[2] Southbound from Manhattan and on the New Jersey side of the river, a different kind of power stirred: Edison's Menlo Park "invention factory"—where the Ohio native intended to develop "a minor invention every ten days and a big thing every six months."[3] He'd started out as a Western Union operator, spending his free time digesting what he could about telegraphy and electricity until, in 1874, he sold the rights to his quadruplex telegraph to Jay Gould (Western Union's rival) for a shocking $30,000. Today, that would equal just over half a million dollars. He reinvested his worth back into Menlo Park lab, soon to be the center of the electric universe. Serious-faced, square of jaw, with eyebrows that curled up like a pair of ragged wings, Edison could thumb his suspenders with dogged determinism. "I haven't failed," he once remarked, "I've just found 10,000 ways that didn't work." If 1st Baron William Armstrong represented the top hat and tails of Victorian gentlemen engineers, Edison offered its thick-booted version, the epitome of American industry and blue-collar resolve. Instead of a wonder-house of working hydro-technology perched on the precipice of a vast country estate (à la Cragside), visitors to Menlo Park would find a rattling, humming machine of industry. On the first floor, visitors encountered "scribes and bookkeepers in one end, and at the other some ten or twelve skillful workers in iron," while the second floor stood ablaze with natural light from wide windows, illuminating "batteries of all descriptions, microscopes, magnifying glasses, crucibles, retorts, an ash-covered forge, and all the apparatus of a chemist."[4] Through the din of metalworking, the dust of shavings, and the stink of battery acid, Edison moved like a benevolent dictator. Problems meant work. Work meant patents. America's greatest engineer was on his way to powering the future. But he wasn't getting there alone.

George Shattuck Morison described the new age as one of collaborators rather than inventors: "The first steps in invention and in new developments are taken by individuals; the best work is done later when the path into which the bold inventor ventured alone is trodden by the crowd who find it their usual course."[5] Like a latter-day Royal Society Newton, Edison collected men. Many of his best associates moved with him from Newark, including Charles Batchelor, a Manchester mechanic with extensive skill in motors, and John Kruesi, the Swiss machinist who "translated" Edison's rough sketches into "high-quality working models."[6] When, on September 16, 1878, the *New York Sun* published the headline of EDISON'S NEWEST MARVEL: SENDING CHEAP LIGHT, HEAT, AND POWER BY ELECTRICITY, the wizard was capitalizing not on his own discoveries, but on the hours of his labor force, the men who would help him bring it to fruition. Edison's vision "had already vaulted beyond mere light bulbs into a glorious and immediate future of electrical grandiosity," explains Jonnes, for "not only did he have in hand a workable incandescent light bulb" (the holy grail of engineers), he planned to "create an entire electrical power system."[7] Of course, Edison had yet to achieve the necessary patents in 1878, and though he'd worked out how to keep an incandescent lamp filament from melting (one of the principal problems), he still had only vision and hope. Even so, he claimed he could, and would, "light the entire lower part of New York city, using a 500-horsepower engine."[8] Like Armstrong and Brunel, Edison had grasped not only what power could do, but what it could be used for.

The *New York Daily Graphic* praised Edison for his practicality in business: "the keynote of his work is commercial utility. He asks himself when a new idea is suggested, 'Will this be valuable from the industrial point of view?'"[9] Will it *turn a profit*? Success followed money, and money followed success. If the world were to be lit by electric light, someone needed to produce it—and to distribute it—and at a price. Investors poured in and the Edison Electric Light Company launched on October 16 of the same year with 3,000 shares. Edison retained 2,500 of them (at an astonishing $250,000) and the remaining 500 went to key investors, including the Vanderbilts.[10] And Edison was good for it—flourishing his new light with gusto. No more of Davy's too-bright arc lights, no more of the flammable gaslight, or the increasingly rare whale oil; Edison's light burned "clear, cold, and beautiful" with the "phosphorescent effulgence of the star."[11] Edison had closed the circuit in a way even Faraday couldn't have imagined (though he read all Faraday's works): he created the light bulbs—which created the need for electricity—which he supplied by his power stations on his lines to his customers. On September 4, 1882, Edison's Pearl Street Station came online, providing direct current (DC) to fifty-nine customers in Manhattan. The reasons for direct current were manifestly simple: there wasn't any other kind. Not then. And until 1888, there could be no doubt that Edison was the world's most formidable engineer, a name known in France and Germany as well as England and the United States. A decade later, however, and it would be Westinghouse, not Edison Electric, powering the astonishing lights, Ferris wheels, show boats, and bulbs at the Chicago World's Fair—and not with DC but with AC (alternating current). That incredible force, a force that changed the entire electrified world, had been made possible not by Edison, but by a quirky Serbian engineer who once worked for him: Nikola Tesla. So how did a man like Tesla, a seer of visions with less grasp of business than even Charles Babbage, eclipse the practical engineer of Menlo Park? Tesla's story begins where many early inventors' stories end; in nearly every way, he lived perpetually in the future, having overshot the science entirely. So brilliant were his plans for use five, ten, twenty years down the line that the actual work of building and making the technology *today* seemed a sort of annoyance, an afterthought, a thing to surmount rather than the thing itself. Elting described this risky behavior in his story of the *Wampanoag*: skip too many steps and you will lose. The path to greatness from Armstrong to Edison built on the tech that came before for a market that was ready to receive it. It sunk Babbage's engine, and it nearly sunk Brunel's big ship. We can usually predict the failure of those inventors who don't align; as I said in chapter 5, history's inventions never jump the rails. The trouble is, nonfiction doesn't have to be plausible . . . and no one predicted Tesla.

Bright eyes, raven hair, fitted suit (the cut and size of which he would wear well into his sixties), Tesla moved like a cat, ate like a lion, and swam like the wood ducks he'd grown up with in lakes and rivers of Serbia. Tesla was a walking mathematical enigma; he counted everything compulsively, and ensured that all his activities (even the laps he swam or meals he ate) were divisible by three. His autobiography reads like *Frankenstein*, and his career zings with Edison-like newspaper headlines. Biographer Marc Seifer took an academic's method to sorting fact from fiction, but even he leaves gaps and makes leaps, revealing a composite being part-hero, part madman, an invalid, a manipulator, and a super-brain. Given his bizarre temperament and habits, it isn't surprising that Tesla failed so often in his life; it's surprising that he ever succeeded at all. And we have to ask why—why did the careful narrative of science and engineering grind its gears in the explosive years between 1880 and 1899, between Armstrong and Tesla, between the Victorian Age and the dawn of a new modern era? Stepping from the train platform amid the smoke and steam and man-made light of Paris, Nikola Tesla walked into the future ready-made for fiction and the myriad films and novels that would resurrect him for the twenty-first century.

## *Changing the Story*

For Tesla, power was life; he was born to claim it as his own. He was also born (so he writes in his autobiography) during the tumult of a violent electrical storm. It reads so much like Victor Frankenstein's family narrative that it's tempting to assume he modeled his own story upon it, and in fact, this wouldn't be the first time Tesla stood accused of copying fiction. His inventions seem sprung from the pages of a nineteenth-century novel, *Vril, the Power of the Coming Race*, though Tesla denied needing any motivation except that of nature herself. At the age of twelve, he saw a picture of Niagara Falls. The impressive sweep, slung round in a horseshoe of thunderous and foaming cascades, sends an average six million cubic feet of water over the crestline every minute. Though a black-and-white image, the falls made a deep impression. Tesla told his uncle then that "one day" he would harness it. A pretty childhood dream, except that, in 1897, Tesla would give the inaugural speech of the Niagara Power Station. The years between would be hailed as the "war of currents," as ready-made for fiction as any wizard's duel: American grit, ingenuity, and bullish pride against a near-clairvoyant visionary from abroad, with his flashes of light and electrical wonders. It's little wonder the two men appear so often in steampunk. A fictional Tesla turns up in Ralph Vaughan's four Sherlock Holmes/H. P. Lovecraft crossovers, *The Adventure of the Ancient Gods* (1990), *Sherlock Holmes in the Dreaming Detective* (1992), "The Adventure of the Laughing Moonbeast" (1992), and *Sherlock Holmes and the Terror Out of Time* (2002). He turns up, too, in Christopher Priest's 1995 novel and later Christopher Nolan's steampunk movie adaptation, *The Prestige*. Edison gets his mentions too. In *Five Fists of Science* by Matt Fraction, Tesla teams up with Mark Twain to fight a dastardly Thomas Edison and his financiers J. P. Morgan and Andrew Carnegie. Tesla's quirks are taken as rote and amplified, while Edison's brash and bullying ways make him ready fodder for evildoing. In *The Lives of Eccentrics* by Hirohiko Araki, Edison is violent and cruel to his workers, while in *Five Fists*, he performs occult human sacrifice. The desire and dread of technological progress shows itself best when fiction and fact collide: light and life, dread tech and death. The real-life men and their real-life contest were not nearly so dark, but they're not as straightforward either.

Tesla arrived in Paris in 1882—a city he would often describe as a wonderland—to work for Edison's man Charles Batchelor in Ivry-sur-Seine (at Edison's Société Industrielle). It would be some time before Tesla met Edison himself, but they were not instant enemies by a long shot. Edison teased Tesla for his manners, calling him "our Parisian"; Tesla writes that the moment was "thrilling," but that he also received his first comeuppance: "I wanted to have my shoes shined, something I considered beneath my dignity. Edison said, 'You will shine the shoes yourself and like it.'"[12] Tesla did exactly as instructed. He would not be treated any differently than Edison's other hardworking engineers, for he permitted no exceptions. But Tesla was (as he repeatedly pointed out) *exceptional.* Tesla describes his childhood home as idyllic, but also shattered by personal tragedy. His autobiography keeps to this type of balance, but it is uneven terrain. He glosses over fundamental years of his life: "I had a complete nervous breakdown and while the malady lasted I observed many phenomena strange and unbelievable,"[13] and then spends pages on details and minutiae of other seemingly more trivial occurrences. He makes lists, citing manias against certain colors, a hatred of earrings, and a deep dislike for the touch of human hair. He had strange powers of eidetic memory, seeing the things he imagined as though they were real, had what we might call out-of-body experiences, and suffered strange flashes of light. Tesla claimed they preceded his discoveries, but he never attributed the psychoneurological disturbances to anything like providence (an interesting point as he was originally destined for the clergy like his father). By his own admission, Tesla had been "almost drowned," "nearly boiled alive," and "just missed being cremated"; he became "entombed, lost and frozen," and escaped "mad dogs, hogs, and other wild animals," and yet he believed in no miracles and accepted nothing but the solid, material world as his foundation. Tesla seemed somewhat miraculous himself, all the same; he could "mold his various creations, and even run and modify them in his mind" before writing up his first blueprint.[14] His litany of bizarre characteristics are given as though already highly familiar to his audience (and they probably were). But he is just as free with accolades for his genius. "We have, undoubtedly, certain finer fibers that enable us to perceive truths when logical deduction, or any other willful effort of the brain is futile," Tesla explained.[15] And it was due, he claimed, to the same principle of energy and power mankind had so long attempted to control.

## The Secret of Life

> *"[T]here is no word in any language I know which is an exact syn-*
> *onym for vril. I should call it electricity, except that it comprehends*
> *in its manifold branches other forces of nature . . . These people*
> *consider that in vril they have arrived at the unity in natural*
> *energetic agencies . . . they can exercise influence over minds, and*
> *bodies animal and vegetable, to an extent not surpassed in the*
> *romances of our mystics."*
>
> —Edward Bulwer-Lytton,
> *Vril: The Power of the Coming Race*

"Of all the forms of nature's immeasurable, all-pervading energy, which ever and ever change and move," Tesla announced, "electricity and magnetism are perhaps the most fascinating."[16] We live within a teeming sea of electric impulses, the hum and beat of their movement within our bodies as well as without. Who among us can readily grasp the idea that our sturdy bones and muscles are vibrating with electrons or spinning with atoms? Recall the first images of the human cell, or of neurons, or any of the other tiny structures of biological life: an endless variety of galaxies are here contained, as awesome as the stars and just as uncountable. For Tesla, the gap between humanity and electricity could be easily closed, and to prove it, he would pass tens of thousands of volts *through his own body*. Ada Lovelace would have loved the show and the meaning it suggested: In an electrified state, Tesla grasped exhausted Crookes tubes and caused them to glow. He used his fingers to light up lightbulbs, bestowing power as if by magic, and with the same magisterial wonder commanded by Armstrong on frostbitten nights at the Literary and Philosophical Society. Onlookers compared him to the archangel, and James O'Neill, his first biographer, compared him to a god—though even "the gods of old, in the wildest imaginings of their worshippers, never undertook such gigantic tasks of world-wide dimension."[17]

Imagine for a moment the whirl and dash of white-hot light. Imagine a mere mortal, sparks bounding from his fingertips in a room made darker by his own luminescence—and imagine a voice, soft and precise and accented, speaking from that fire as if from the great beyond: "We are whirling through endless space, with inconceivable speed [. . .] all around us everything is spinning, everything is moving, everywhere is energy."[18] Contemporary accounts describe a scene of wonders that can scarcely be imagined, lamps burning with "magnificent colors" among sheets of light and directed beams. And in the midst of them was Tesla, a genius, a madman, firing up his great coils with the sound of thunderbolts, and claiming that he, like Frankenstein, could endow a mass of shapeless wire with the essence of *life*.[19]

Tesla was not the first to describe the life force of electricity, of course, and electric therapy (as rather horrid quack devices) had already been employed for the better part of a century. For those practitioners, the emphasis had been upon the *shock*. Like so much early medicine, the cure came pretty near to killing you. And life and death did seem to hang in the balance. By mid-century, charlatans of the life spark, animal electricity, and a new foray into "magnetism" tacked up shop signs and spread tents in the malls, and a person of good breeding could get herself invited to a séance. Alison Winter describes a "tea party" in November 1844, where a late-arriving guest arrives to find her hostess lying cold and lifeless on the reclining sofa. The other invitees gather around in awe and fear as a "magnetizer" with "dark, animal eyes" hovered over the entranced woman: the power to hold her in such a state, he claimed, owed to his "moral and intellectual superiority."[20] The guest challenged the hypnotist, who took her hand in his until she felt a current run through it, an "electric" feeling of shock as from a galvanic machine. The experience was known as *mesmerism*, but it didn't originate with the Victorians. Franz Anton Mesmer first produced the effect in *1774*, at the height of the electrical craze and contemporary with the kites and keys and Leiden jars of Benjamin Franklin. Mesmer claimed that the secret to health and even to life itself lay with *animal magnetism*. Each body produced flow, a concept that had gained ground with the new ideas of "flowing" electricity. To Mesmer, illness was a block in this flow that magnetism (the magnetic pull of the practitioner upon his patient) could release. In other words, *mesmerism* was *science*, and predicated on the same principles that caused a frog's leg to jump with touched with copper wire, or which caused static to form on the hand that touched a Hauksbee machine. But Mesmer's work did not get far in his own time; an investigation appointed five commissioners, one of which was Ben Franklin himself. The resulting conclusion: there could be no evidence of animal magnetism. Mesmer was branded a charlatan and driven into exile. So why was a tea-time guest subjected to mesmerism in 1844? Engineers like Edison and Tesla had provided new material, and with each discovery the race was on to find it a home to live in. Electricity was real—why not stranger things? It's a question worth revisiting, and one that would later entrance an aging Arthur Conan Doyle. But first, to the scientists: who would be the first to experiment with electricity's mysterious force?

"My whole scientific education," said English chemist and physicist William Crookes, "has been one long lesson in the exactness of observation, and I wish it to be distinctly understood that this firm conviction [in spiritualism] is the result of most careful investigation."[21] Crookes's work in spectroscopy earned him the respect of the age. He invented the Crookes tubes, glass tubes with vacuum to test electrical discharge, instrumental to the first x-rays and also to Tesla's experiments. A gentleman scientist, white bearded and sporting a comically waxed mustache, Crookes joined the Royal Society, rubbed shoulders with Lord Kelvin, and carried with him a gravitas of credit. And yet, he systematically investigated and supported the claims of leading spiritualists, including Daniel Dunglas Home, Kate Fox, and Florence Cook.[22] Crookes published his own *Researches in the Phenomena of Spiritualism* in 1874, claiming to have witnessed "the movement of material substances, and the production of sounds resembling electric discharges" occurring during spiritualist séances and inexplicable by known physical laws.[23] But "spiritualist" activity did not necessarily suggest paranormal intervention—what if, suggested journalist and medical man Andrew Wynter, technologies like the electric telegraph acted as "spirits" to "carry our thought with the speed of thought to the uttermost ends of the earth"?[24] What if electricity, the spirit, and the body were all one, and the telegraph system "like the great nerves of the human body, unit[ed] in living sympathy all the far-scattered children of men."[25] In other words, what if electricity were the very secret of life and power itself? The answer, for some, turned up first in fiction.

Sir Edward Bulwer-Lytton wrote *Vril, The Power of the Coming Race*, in 1871. Though few recognize his name now, he had a popularity second only to Dickens, and his adventure tales made deep impressions on the reading public. *Vril* begins with a mild-mannered narrator talked into joining an expedition into the heart of a chasm (not so different from Verne's *Journey to the Center of the Earth*). An accident leaves him stranded and alone, and he follows tunnels into a strange and alien world—the home of a master race called the Vril-ya. Though no sun penetrated the endless cavern, the sweeping terrain "was bright and warm as an Italian landscape at noon [. . .] I could distinguish at a distance, whether on the banks of the lake or rivulet, [. . .] embedded amidst the vegetation, buildings that must surely be the homes of men."[26] Like the racing narratives of eighteenth-century explorers in Jamaica or Tahiti, he finds an unknown paradise. Like the crews that imagined worlds beyond the ice, he discovers a race of people unlike other men—but Bulwer-Lytton's underworld is not a step backward to a more primitive society. Instead, the Vril-ya represents an astonishing leap forward. They could burrow "through the most solid substances, and open valleys for culture through the rocks of their subterranean wilderness," and "extract the light which supplies their lamps."[27] They keep no slaves, but are served by strange automatons "so ingenious [. . .] that they actually seem gifted with reason. It was scarcely possible to distinguish the figures I beheld [. . .] from human forms endowed with thought."[28] Everything from vast engines to earth-moving machines operate without the interference or guidance of pilot and driver, all powered entirely by a mysterious substance called *vril*. Just fiction, of course. Except two decades later, Tesla would unveil wireless vehicles, unmanned boats that operated at his command. Rumors circulated. Suspicions abound. Did Tesla take his ideas from Bulwer-Lytton's fiction? Or was vril never a fiction to begin with?

The idea enticed like a magic spell, a new alchemy of power. Vril offered just enough possibility (and profitability) that some so-called engineers took the part of medical charlatans, offering false science at the expense of a willing public. The worst of these offenders might be John Ernst Worrell Keely. He claimed to have invented a perpetual motion machine, a motor which proved the existence of this "mysterious force of the universe."[29] Using senseless jargon, a showman's platform, and something akin to religious fervor, Keely would commune with this "essence" to run the motor, which would release steam as evidence of its operation. Eventually arrested, Keely divulged his secret: A connecting copper tube moved the motor on his signal. So much for magic. But, then, electrical science sounded almost as far-fetched. Edison declared he'd light up lower New York: Tesla said he would put "100,000 horse-power on a wire."[30] Edison wanted to sell affordable power: Tesla claimed mankind could access the power driving the universe for free, powering lights and small crafts without a single connecting thread. When Tesla unveiled his first "teleautomatics" machine, a boat controlled by remote radio transmitter, at the Electrical Exposition in

Madison Square Garden, he caused a stir. Even a hundred years after Davy's arc light exhibition, everyone still loved a show.

We have all seen and probably played with Christmas toys of one variety or another, robots and cars that trundle along on short-lived batteries and controlled by invisible waves from our remote antennae. Most of us probably never thought to question the whys and wherefores, but imagine such a thing at its first debut, as amazing in its way as Jaquet-Droz's writing boy. Taken to the next level, Tesla's boat had been equipped with the ability to communicate and even to answer questions through signals, just like a trained circus counting horse. In reality, the same radio transmission that controlled the boat's movement in the water also allowed Tesla to give responses on its behalf, but the public called the boat "magical," and Tesla let them. "You see with what ease the Vril-ya destroy their enemies?" Bulwer-Lytton's novel asks. Tesla's model could fire explosives by remote control, and a similarly radio-guided torpedo offered the possibility of disaster at a distance, not unlike Armstrong's Big Gun.[31] And also like it, those who did see its possibilities worried about an era where "quarrels will be settled in view of the terror of the cataclysms promised by science."[32] That was partly the point of *Vril*'s "utopic vision"; among the Vril-ya, the "art of destruction" through death-ray type weapons had surpassed everything, and so annulled all "superiority in numbers, discipline, or military skill."[33] As a result, the "age of war was therefore gone." If this seems illogical, consider: WWI was called the "war to end all wars." The Cold War, too, with its proliferation of nuclear weapons no one wanted to use, suggested that humankind had learned from the horror of destruction not to use their powers for evil. In theory. The trouble, of course, has to do with practice. I mentioned that Tesla and Twain team up in *Five Fists of Science*; in the novel, the two of them build giant automatons against which "fleets and armies would be helpless," not to destroy the world but to save it. . . . In history, Mark Twain did in fact approach Tesla. He urged him to sell the patents of the remote control boats to all countries. That way, with such a level playing field, war would have no place. We can perhaps be pleased that Tesla never did; the trouble isn't with Bulwer-Lytton's technological premise, but with the length of human memory. The power of vril, unleashed into military designs, would not guard against future generations who lived to forget the dead and the dying. And perhaps this, as much as the sheer implausibility, is what caused the military to balk. Guns to bigger guns they could understand. Death by magic rays and invisible ether was not only harder to conceptualize, it had no precedent. Then again, neither did AC power systems when Edison opened his Pearl Street Station. And Tesla would work on the concept of death rays well into his waning years.

### Birth of a Myth-Maker

To understand how Tesla arrives at his vril-like teleautomatics, we have to start further back. Like Frankenstein (and, by extension, its author), Tesla found the wonders of nature utterly fascinating. Also like him, Tesla wasn't originally in pursuit of destructive creations, but in harnessing lightning and taming it. Tesla wrote, "if only we could produce electric effects of the required quality, this whole planet and the conditions of existence on it could be transformed." His aim was no less than godlike; "The sun raises the water of the oceans and winds drive it to distant regions where it remains in a state of most delicate balance. If it were in our power to upset it when and wherever desired, this mighty life-sustaining stream could be at will controlled."[34] What might men do if they harnessed a power as great as the sun? Might we, like Victor Frankenstein, discover the secret of life itself? The answer certainly seemed to be electricity—but even caught "in a bottle" use of so sublime a substance had long eluded man's control.

"Practical" electricity simply didn't exist during Tesla's school years; Michael Faraday had long since demonstrated that electrical current *alternates* in its natural state, that is, the flow of electric charge reverses direction, but no one had sorted out how to harness it for everyday uses. Marc Seifer compares the electric flow to water in a river—if the river keeps changing direction on you, running a water wheel would seem impossible or at least impractical. The Gramme dynamo (or motor) used a commutator, or series of wire brushes, to translate the current into a single direction for use—the *direct current* that later powered Edison's light. But by forcing the current to flow only one way, the commutator also wasted energy by encumbering the motor (and throwing sparks in the process).[35] The spitting, whirring electrical motors simply didn't work any other way; they never had, and, as Tesla's professor Poeschl explained, *"they never will."* Wasn't it worth trying to capture AC instead, Tesla asked? "Mr. Tesla may do many things," Poeschl chided, "but this he cannot accomplish. His plan"—like Keely's—"is simply a perpetual motion scheme."[36] But, unlike Keely, Tesla was no charlatan, and he would not cease until he'd proven them all wrong. "With me it was a sacred vow," he claimed, "a question of life and death."[37]

Young, untried, but certain, the young Tesla spent every spare moment (not in classes) on his designs. He did not eat. He did not sleep. For months, he groped his weary way through corridors of thought and could focus on nothing else. Imagine a bright mind so fixated on a single pinprick of light—endlessly winding around and around the same problem. His closest friends worried about his sanity—with good reason. As early as 1835, James Cowles Prichard had identified "monomania" as a type of madness, a partial insanity "in which the understanding is disordered under the influence of some particular illusion, referring to one subject, and involving one train of ideas."[38] In 1845, the English translation* of Jean-Étienne Esquirol's treatise added that there "has been no advancement in the sciences, no invention in the arts, nor any important innovation, which has not served as a cause of monomania."[39] Granted, the prognosis might apply to most of the genius and inventive minds we've so far encountered; in fiction as well

---

\*     In French, from 1820, *Il n'est point de progrès dans les sciences, d'invention dans les arts, d'innovation importante qui n'aient servi de causes à la monomanie, ou qui ne lui aient prêté leur caractère.*

as fact, it is unwise to tell a genius (even a mad one) that he *can't*. But the strain did result in nervous and physical collapse.

The AC discovery makes up the better part of Tesla's autobiography, his heroic epic at climax, life hanging in the balance, heart racing 250 beats per minute: "I could hear the ticking of a watch with three rooms between [. . .] A fly alighting on a table in the room would cause a dull thud in my ear. A carriage passing at a distance of a few miles fairly shook my whole body. The whistle of a locomotive twenty or thirty miles away made the bench or chair on which I sat vibrate so strongly that the pain was unbearable."[40] Emaciated, trembling, assaulted by his own sense of thwarted becoming, Tesla did not expect to recover. His friend and fellow engineer Anthony Szigeti carried him daily into open air that he might have the benefit and balm of nature—where at last, the flash of inspiration came. Watching the rays of a fading sun, Tesla recited Goethe, whose *Faust* he had memorized by heart. The words are of failure, of loss.* But, as soon as they left his lips, Tesla writes, "the idea came like a flash of lightning [. . .] I drew with a stick on the sand the diagrams shown six years later in my address before the American Institute of Electrical Engineers." Tesla had invented, in that instant, the first AC motors with rotating magnetic fields.

Whether it be truth or the exaggeration of a good storyteller, it's impossible not to hear notes of Victor Frankenstein's triumph in Tesla's words: "A thousand secrets of nature which I might have stumbled upon accidentally I would have given for that one which I had wrested from her against all odds and at the peril of my existence."[41] His diagrams awed other engineers,

---

* *Sie rückt und weicht, der Tag ist überlebt, Dort eilt sie hin und fördert neues Leben. Oh, das skein Flügel mich vom Boden hebt Ihr nach und immer nach zu streben [. . .] Ein Schöner Traum indessen sie entweicht, Ach, zu des Geistes Flügeln wird so leicht Kein körperlicher Flügel sich gesellen!*
"The glow retreats, done in the day of toil;/It yonder hastes, new fields of life exploring;/ Ah, that no wing can lift me from the soil,/Upon its track to follow, follow soaring! [. . .] A glorious dream! Though now the glories fade./Alas! The wings that lift the mind no aid/ Of wings to lift the body can bequeath me."

not with complexity but with their simplicity. Unlike those before him, Tesla used two circuits instead of one, generating dual currents that were 90 degrees out of phase with each other. "The net effect," writes Seifer, "was that a receiving magnet (or motor armature), by means of induction, would rotate in space and thereby continually attract a steady stream of electrons whether or not the charge was positive or negative."[42] If we return to the water wheel problem on a river whose current keeps changing direction—the trick was to have 1) a second stream of "water" and 2) a wheel that could rotate in space in either direction. It wouldn't work on a river, but it did work in Tesla's motor. The lightning strike had revealed what more than four years of monomaniacal thinking had not produced: Poeschl had been wrong—harnessing AC was not impossible, and more importantly, Mr. Tesla *could* do it. "Isn't it beautiful?" he asks Szigeti, "Isn't it Sublime? [. . .] I must return to work and build the motor so I can give it to the world. No more will men be slaves to hard tasks. My motor will set them free, it will do the work of the world."[43] He needed only to find those who would support it—and finally the story catches up with Edison. Tesla went to see the Wizard.

When the reporters saw Edison's first bulb with its platinum filament, they said it was a mechanism "so simple and so perfect that it explained itself." In reality, the filament was inferior and burned out within an hour; the reveal had been staged to please the press and gain the support of investors. It would be months of sleepless labor before the Menlo Park lab developed the carbon filaments that ultimately made the lightbulb successful. With Tesla, however, the concept of an alternating, polyphaser motor truly operated on the engineering mind as a sudden, brilliant, and obvious concept. His design—modeled above in Seifer's metaphor—was so easy to explain, it seemed intuitive to other engineers. Surely anyone (of similar training) could have come up with it, they argued, and Tesla ultimately went to court to protect his priority. Judge Townsend ruled in Tesla's favor, however, calling it the "doctrine of the obvious": "The apparent simplicity of a new device often leads an inexperienced person to think that it would have occurred to anyone [. . .], but the decisive answer is that with dozens and perhaps hundreds of others laboring in the same field, *it had never occurred to anyone before*" [*Potta v. Creager*, 155 US 597, author's italics].[44] Success, but short lived; the world did not jump at Tesla's new idea—partly because they had no need, no market. Edison's system worked, and Tesla's was a "quantum leap" to grasp, a *Wampanoag* of an idea. But surely a man of Edison's ilk would listen, Tesla reasoned, and one late night over dinner, he explained his system. Edison's power stations could only reach a mile in any direction; DC simply wasn't strong enough. But now, with the ability to run motors and engines on AC current (the same way Davy's arc light had been powered, but with far more security), Edison's plants would power whole cities. Even the nation. The AC motor would be superior, Edison's company would power it: the world was their oyster. But Edison would not have it.

Tesla relays to his first biographer, O'Neil, that Edison refused on two grounds. First, he had already established himself in DC and saw no future in AC, and second, AC (with its high voltage) was dangerous in the extreme. He wanted nothing to do with Tesla's vision. It reflects the same problem Armstrong and Brunel encountered; if you would succeed, you needed a public and investors too. But Tesla could not wait and hope, as Brunel had done with his enormous *Great Eastern* ship. It was the future he was after, and he needed to find a risk-taker. Perhaps not surprisingly, he found his man among Edison's rivals: George Westinghouse. Fiction gets it wrong; the "war of currents" didn't actually play out between Edison and Tesla, but between Edison Electric and Westinghouse Electric, to whom Tesla sold his patents. The real fight wasn't between engineers, but among them—not between old and new technology, but between new and newer technology. Edison was a practical engineer . . . but he was also *wrong*. And for once, a quantum leap made the best sense.

If Edison traded upon the engineers in his invention factory, Tesla traded upon the future—and so, in his way, did Westinghouse. An American Armstrong; he quadrupled the worth of his company from $800,000 in 1887 to $3 million in 1888, despite being embroiled in legal battles with Edison over lightbulbs and patents.[45] The deal made between Tesla and Westinghouse favored Tesla and provided him with ready money and incredible royalties. But by effectively tying himself to Westinghouse Electric, Tesla became the (perhaps inadvertent) enemy of Edison in a struggle over who would control the power grids. Edison's first argument, that there wasn't a future in AC, didn't offer much to the unlearned public. But the second argument moved hearts. By claiming that AC's high voltage and extraordinary power meant certain death, Edison tapped into a very real fear; after all, plenty of accidents had already occurred. He opted for propaganda to show the danger of AC current—electrocuting dogs and even elephants in public squares, and most (in)famously, promoting the use of electricity as a means of capital punishment. H. P. Brown manufactured the first electric chair and sold it to prisons for $1,600, and the first test subject would be William Kemmler.[46] Edison relished the idea; a public display of electricity's evil work on the human body—surely, AC would lose all credit with customers.

Kemmler had been convicted in 1889 for the murder of his common-law wife . . . with a *hatchet*. On August 6, 1890, he was strapped, blindfolded, into a chair for electrocution by AC current. But things did not go as planned. "To the horror of all present," remarked a witness, the electrocution did not provide a swift end; in a scene that rivals Aldini's experiments and Mary Shelley's *Frankenstein*, "the chest began to heave, foam issued from the mouth, and the man gave every evidence of reviving."[47] It took a second shock of 1,000 volts to finish the job, by which time Kemmler's head caught on fire and the rear electrode burned through to the spine. With some irony, a deeply distressed George Westinghouse suggested they might have done better by taking up Kemmler's own method and using an axe. Either way, the event became a public relations nightmare, and nearly sunk Westinghouse (and Tesla, and AC by extension). But Tesla remained undaunted. After all, Edison wasn't the only showman—or the only one willing to bend the truth.

One of Tesla's famous photographs depicts him sitting quietly, one hand to his temple, while arcs of live fire dance all around his laboratory. The photo is a fiction. Tesla had the photographer take double exposure to make him appear to be in the room at the same time as the volts, a kind of "spirit photography" as popular at the time as séances and mesmerism [Fig. 14]. Most people didn't know the camera could *lie* (an uncritical view that would lead to Arthur Conan Doyle's support of the Cottingley Fairies, among other things). Why would Tesla, who already demonstrated live that electricity could pass through his body, take the further step of "faking" a relaxing moment in the presence of AC current? It would be one of Tesla's early, but by no means first, attempts at fictionalizing fact, at mythmaking. And he had very good reasons. He was not, like Frankenstein, interested in bringing the dead to life—he was interested in life as an eternal and exacting force, electricity in the ether, and the possibility that he would harness it where all other men had failed. Westinghouse feared the public too much to continue work on Tesla's big ideas, so Tesla traded again. This time, he bartered away his royalties to ensure the project would move forward despite the backlash. And, through presentations, experiments, and photographs, he would woo the public's favor himself.

Westinghouse used transformers at a frequency of 133 cycles and did not want to depart from their standard apparatus. Instead, he wanted Tesla to adapt his polyphase designs to fit them. The lost time meant Tesla couldn't commit full resources to discovering how far AC could travel and with what efficiency (a necessary step for bringing the system and motors together). He didn't have an invention factory with hundreds of willing workers . . . but there were other ways of ensuring the ideas stayed public. Tesla gave a lecture, and later that year, German engineers followed his methods and sent 190 horsepower from a waterfall to a factory 112 miles away in a stunning light display. (Plenty of headlines, lots of exclamation points!) But the Germans didn't invent it; they merely copied from Tesla's patents—most probably, Marc Seifer suggests, with Tesla's full knowledge and blessing. It was within his rights to sell usage of the patents, and what better way to prove that he'd been right? The incident would become highly significant in the battle for Niagara power, but though Westinghouse would finally agree to do it Tesla's way, Tesla himself doesn't get all the credit. Instead, the company would promote a relative unknown named Benjamin Lamme. A new hire, he had less baggage and would be less "trouble."* Edison was never Tesla's greatest rival. His real problems weren't caused by individual men so much as corporations—and perhaps his own vanity concerning what was or was not beneath him. He'd made enemies among other engineers, and also among investors; he had grown poor by refusing to accept a salary that would put him on par with "regular" workers. It was one of Edison's first criticisms of Tesla and probably the one thing that haunted his genius most. Tesla didn't value allies. He wanted to be Newton—he would stand (or fall) alone, a mad scientist indeed.

## *Troubling the Future*

How did AC, branded as a vicious killer by its enemies and an expensive retrofit by its supporters, finally win over DC? For one thing, it *worked*. Once Tesla's AC motor proved the efficiency of alternating current for practical uses, and in particular because it garnered so much attention not only through Tesla's presentations but also through court cases, it was only a matter of time before other engineers put it to good use. Edison's public campaign against it probably did more to light the fires of curiosity in a rising generation of electricians than it did to temporarily hurt business. And as those new thinkers came of age and filtered into the workplaces, not only did AC become impossible to contain, but it became harder and harder to defend those hard-won patents. By the 1880s, Edison's career in electric light winked out, and he conceded defeat, writing in a letter that

---

* Marc Seifer describes the stalemate at Westinghouse; Tesla had said all along that the motor could not be adapted, but to admit it would be costly in terms of money and pride. A coup of sorts resulted in promoting Lamme as the inventor of the "solution" rather than the motor (protecting patents), just one more way that Tesla's name was pushed further from the center of credit.

his "usefulness" to the power industry had gone, and that he sought gradual retirement from the business (though not from work; he continued active and inventive all his life). General Electric had subsumed Edison's company, which gave them the right to make Edison's bulbs—but in the new AC world, Westinghouse generated all the power. There must be a way around Westinghouse AC, GE argued. There was just one problem. Tesla still held all of the relevant patents to the constituent part. As Seifer puts it, "Quite simply, there *was* no other system."[48] And yet, in a curious twist, the years that followed saw both the "wizards" ousted from the powers they originally commanded. Edison lost to Tesla, and Tesla lost to GE and a political exile named Charles Proteus Steinmetz.

Almost the physical opposite of Tesla, Steinmetz was small of stature (at four feet) and suffered a hunched back and a limp, consequence of one leg being shorter than the other. He was, however, an incredible mind. The solution for GE was to erase Tesla's involvement entirely, and they played a long game to try to achieve it, starting with Steinmetz. Up until this time, Tesla's own work (edited by T. C. Martin) offered the most comprehensive guide for young engineers: *The Inventions, Researches, and Writings of Nikola Tesla*. Like his later autobiography, even the title made sure to privilege Tesla's reputation. By 1897, they began promoting Steinmetz's competing volume, *Theory and Calculation of Alternating Current Phenomena*, which described the entire system without mentioning Tesla's name—not even once. Given that young engineers had little grasp of the history, and Westinghouse's own frustrations with Tesla kept them from leaping to the defense, a whole generation absorbed a highly edited history. "It was quite common in the later years," Seifer remarks, "for engineers to obtain degrees, study AC and even write textbooks" and never learn Tesla's name.[49] Steinmetz spoke about AC in the very way that those early opponents spoke of the motor: it was an obvious solution, one that occurred to many engineers, and, after all, its success demonstrates it as a foregone conclusion. Tesla was *not* the wizard of AC current; he was merely one more engineer who once worked for the electrical giants, GE and Westinghouse (names that still remain at the forefront of power today).

It should have crippled Tesla's reputation, assigning him a place among the forgotten ones. But Tesla was not just *any* inventor, and never worked on only one thing or even in only one area of things. Lectures at Columbia, meetings in London with the Royal Society, rubbing elbows with elites that Edison himself would scarcely have a chance at (including Crookes and Lord Kelvin), Telsa might not have has as many allies as Edison, but he knew where to make friends and how to impress. He also knew how to speak multiple languages, not just national tongues, but also the language of both the educated elite and the practical engineer.

When Tesla commanded electricity, he explained the process all the while: "the sparks cease when the metal in my hand touches the wire. My arm is now traversed by a powerful electric current, vibrating at about the rate of one million times a second."[50] All around, says Tesla, "the electrostatic force makes itself felt [. . .] hammering violently against my body [. . .] I can make these streams of light visible to all."[51] He'd realized that alternating currents of up to 10kHz could pass through the body before anyone recognized the sensation of "shock." He'd also learned just how much a body might take of the electric fire. Using high-frequency currents, Tesla discovered that living tissue could safely be heated and that the electricity from his oscillators could work changes in blood pressure.[52] He even delivered his paper to the American Electro-Therapeutic Association in 1898, when a series of electrotherapy devices were being reviewed for practicality in medicine (as opposed to the far more dangerous quack devices like the electrified bath of mid-century which was disastrous for somewhat obvious reasons). Tesla did nothing in a corner; he used the press, headlines, and lectures to make himself the center of the electrical universe.

By 1893, Tesla had become the darling of the press, announcing in front of his greatest peers, "Is there, I ask, can there be, a more interesting study than that of alternating current?" He had become its master and arbiter, so much the master that when fire destroyed his New York lab, the world collectively gasped—Charles Dana, a revered newspaper editor, proclaimed: "The destruction of Nikola Tesla's workshop with its wonderful contents is something more than a private calamity. It is a misfortune to the whole world."[53] Westinghouse would go on to win the Chicago World's Fair contract, and Tesla would go on to give his remarkable speech at Niagara Falls—a speech where he proclaims the massive station as only the beginning, only a step on a broader and more worthy path: "We have a greater task to fulfill, to evolve means for obtaining energy from stores which are forever inexhaustible, to perfect methods which do not imply consumption and waste of any material whatever [. . .] I shall see the fulfillment of my fondest dreams; namely, the transmission of power from station to station without the employment of any connecting wire."[54] We've arrived at Tesla's great achievement, the height of a career still in the making and on the rise. Standing before crowds and the power of a thundering cataract in harness, Tesla's toast commemorated the end of the war of electrical currents, all the devices and maneuvers of his adversaries notwithstanding.

A new wizard had taken the stage. His success with AC propelled him into the circle of the scientists. It was Lord Kelvin, upon being presented with one of Tesla's coils, who suggested he go further—that he "concentrate upon one big idea." As Seifer explains, such a charge, coming from so great a hero of science (Kelvin worked out the wavelength of light and atomic weights of metals, among other things), shook Tesla with a sense of destiny. He wanted to do something greater than AC power, greater even than the Niagara power station. The future, he thought, must be in wireless—in his eventual "telautomatics," his version of *vril*, and with weapons of mass destruction that he never built.

Tesla's days as the premier electrical wizard may have been longer than Edison's, but they were still numbered. He'd tasted success, but he longed for an ideal future. In the midst of his Niagara speech, Tesla told the gathered audience that it means little in comparison to a future of *free* power. He was necessarily interrupted by the businessmen who found that idea both ludicrous and unprofitable—what did Tesla mean by saying men could pluck energy from the air as easily as plants take nourishment from the sun? "Electric energy can be applied to bicycles, carriages, and vehicles of all sorts," Tesla explained in an interview to *Electrical World*, "so cheap that any man in ordinary circumstances can own a boat and propel it by these means." After all, "it would be a gloomy prospect, indeed, for the world if we did not think that this great power will be used to the advantage of the vast majority of the human race."[55] But the careful control of his early narrative—the successful story that ultimately brings down Edison and converts the raging Niagara of his childhood vision into cheap and reliable power—fractures in Tesla's late career. Inventions chase inventions; he's on to the next before the last one is complete, to the frustration and alienation of his investors, supporters, and friends. Adam Steltzner, a Delta Systems engineer from the Jet Propulsion Laboratory, claimed that "as engineers, we don't get paid to do things right, we get paid to do them just right enough."[56] Or as a colleague of mine has put it, "don't aim for the stars, aim for the door."* Tesla's greatness and also his weakness lie in his far-flung desires and otherworldly designs. The success of his first quantum leap, that of AC power, had in fact been made possible by the Wizard of Menlo Park, the businessman who had laid the wires and made the bulbs and, by extension, created a market for power. DC was "just right enough." AC led to a future none had foreseen but Tesla, but it was still a future that made sense to them, a future of business and power and money and industrial strength. The practical was never part of Tesla's own narrative, and certainly not the crown of his achievements. But his shadow is long, and suggests in its shapeless dark greater possibilities than light bulbs and electric motors. There is a sense of singular greatness, of a height and breadth of achievement that pushes the limits of a single lifetime, and it makes Tesla's story larger than his life—a great woven tapestry of fact and fiction that seduces with its very ambiguity. He knew, or thought he knew, what power was for: wireless technology, cheap manual labor, remote control planes, radio transmitter torpedoes, and telecommunications. Sometimes, in aiming for

---

\* With thanks to Sean Gordon.

the stars, you accidentally reach them, but the future (and in fact, the *present*) was far darker than Tesla—or George Shattuck Morison, imagined. In that, Tesla is perhaps no engineer at all, but a harbinger of a future where we, too, take leaps in the dark without first taking a sounding, a Babbage, and not an Armstrong at all.

"Oh, we marched and ranted, for the rights of labor and such—fine talk, girl!" says Mick to Sybil in *The Difference Engine*. "But Lord Charles Babbage made blueprints while we made pamphlets. And his blueprints built this world!" Sybil feels the mechanism beneath her feet: "the digging went on, this and every night, for she'd heard their great machine huffing, as she'd stood by Mick on the Whitechapel pavement; they worked unceasingly, the excavators, boring newer, deeper lines now, down below the tangle of sewers and gas-pipes and bricked-over rivers." Can we imagine a world such as this? We can—we can because we live in it, in great craters of cities over great canyons of steel pipe and turbines and twisting tunnels of electricity and conductors. The danger threatens when change comes too fast. Gibson and Sterling's fiction is exactly this kind of collision, a point at which man invents faster than he can accommodate. But we are always, already, in just such a position. Newton and his ilk engaged science to fight our dread of chaos and destruction; by the time of Tesla, scientific and technological advance threatened what we had hoped to thwart or control. The real Babbage had complained of the decline of science; he stood at a crossroads rather than a new epoch, and his work borrowed from John Herschel the language of the cosmos: where were the bright stars? Where were those whose light offered the possibility of great nebula, clearly visible to distant nations, a beacon, a guide into the future of ideas? He could not pursue his original grand plans of the sort that drives *The Difference Engine*'s plot. But the point had tipped, the spark ignited—and as George Morison so aptly summarized, "The old and the new cannot exist together. It is hard to realize how rapidly the appearance of the whole earth may change." The explorers, the captains of sailing ships, steamships, and starships, have the comfortable fiction that the difference is always "out there." Home remains familiar and safe. But what about the changes right under your feet, the ones that destroy the town you always knew, raze the house you were brought up in? "The tendency is apparently involuntary and immediately to protect oneself against the shock of change," writes Elting of his uncle's work. It strikes again at the same old bell: the advent of automatons riddled us with uncanny dread; the age of machinery threatened to fit men to the machines rather than machines fit to serve men—and the possibilities of the difference engine find their most dreadful expression in our fears of inanimate sentience. *The Difference Engine* ends in a single line—but one we can imagine thundering in the cacophony of waking. The machine knows itself. "I see: I see, I see," until it concludes only "I!" It's a sentiment we feel—the uncanny and dreadful, shot through with something like black magic and thunder. In life, both Babbage and Tesla would be forgotten (at least temporarily) by engineers. But they would be remembered in fiction.

In the novel and film *The Prestige*, Tesla builds a "machine" for Mr. Angier,* only to encourage him to destroy it: "Such a thing will bring you only misery."[57] In this story, the genius has become the mad scientist, ill obsessed, guilty of hubris, possessed of the world's greatest mind (and possibly just possessed). "Exact science, Mr. Angier, is not an exact science," Tesla explains. Anything is possible. Even when it is not advisable. "The truly extraordinary is not permitted in science and industry," the fictional Tesla claims, but he also adds, "I followed [my obsessions] too long. I'm their slave . . . and one day they'll choose to destroy me."[58] That complaint, at least, is real. The historical Tesla has as much bitter invective: "I had to cut the path myself, and my hands are still sore [. . .] would you mind telling a reason why this advance should not stand worthily beside the discoveries of Copernicus?"[59] He would try, to the day of his death, to bring about his dream of wireless electricity, but it never came to be. He would attempt

---

\* The machine is meant to transport the user from one place to another; instead it transports a copy of whomever uses it. Mr. Angier spends the last part of the story repeatedly shooting or drowning "himself."

to build death rays, and though unsuccessful, the imagined devastation caused hearts to shudder. The government seized and shut down Tesla's fifteen-story transmission tower; it posed a security risk, they claimed, as the Second World War thundered into being on land, sea, and air. Ruptured becoming thwarted plans. But what would it mean if Tesla had succeeded? Ultimate weapons? Machines of destruction? Upon Tesla's death in 1943, the FBI seized his notebooks and papers, fueling skepticism and rumor about what they contained. Possibly nothing important could be gleaned from them. But the future roared on with plenty of firepower nonetheless, and the atom bomb would be dropped on Hiroshima and Nagasaki two years later. By these technological means, says Elting, we have "moved ourselves from a position of terrifying ignorance and dependence, frail beings in the hands of an angry God, to a place of knowledge and power" in which we become that god, the supreme measure of all things. "This too," he says, "has its terrifying implications."[60] We began with a book, a warning, and the Red Queen. The book proclaimed the New Epoch would wash away our troubles; the warning suggested it would break us in the process . . . And the Red Queen stands at the crossroads of an unruly clockwork of *what-ifs* and *how nows*, of machines that didn't get built and stories not often told. There are no fictions about William Armstrong, some few about Edison, and fewer that feature the likes of Brunel. Success stories of "practical" engineering, they offered up a future of progress. But fiction chose Babbage and Tesla to resurrect. In their struggles, even in their failures, they offer something other, something darker, something brought to life like Frankenstein's monster with the stitches still showing. But what if I told you that this dark un-history was no fiction? What if the steampunk otherworld of *Whitechapel Gods* had been bubbling away all the time in a corner, spiking the air with gunpowder, acid, explosives, and dread? Rotate the dial and turn the gears back we've seen through the eyes of the inventors, engineers, and scientists. Now, we'll set out for the back streets and mine shafts, the upturned world of digger, factory boy, match girl, and urchin on the crusted streets of London at the dawn of the (same) steam-powered age.

# PART FOUR

"The invention of the boat was the invention of shipwrecks. The invention of the steam engine and the locomotive was the invention of derailments. The invention of the highway was the invention of three hundred cars colliding in five minutes. The invention of the airplane was the invention of the plane crash. I believe that from now on, if we wish to continue with technology (and I don't think there will be a Neolithic regression), we must think about both the substance and the accident—substance being both the object and the accident. The negative side of technology and speed was censored. The technicians, by becoming technocrats tended to positivize the object and say, 'I'm hiding it; I'm not showing it.'"

—Paul Virilio, *Pure War*, 1997

## SEVEN

# A Wrench in the Age of Machinery

et's return to 1838, the year of Queen Victoria's coronation, and that legendary meeting of the British Association in Newcastle. It was the beginning of Armstrong's rise to glory—the company he would seek after for the rest of his life, and the first sparks of his future inventions. But the meeting, in the eyes of local people, offered little more than a "vanity fair," a "philosophical traveling circus" of celebrity scientists and their coterie of admiring groupies.[1] Journalist Harriet Martineau left the fanfare in disgust, rather than adulation; "What I saw of that meeting convinced me of the justice, in the main, of Carlyle's sarcasms," she wrote in her biography. "Sadly spoiled [. . .] by the conceit of third-rate men with their specialties, the tiresome talk of one-idea men."[2] She referred to Thomas Carlyle (a satirist and philosopher); he would go on to attack the British Association in his *Fraser's Magazine*, leveling his invective against "all this squabbling" about discoveries they "pretend accurately to determine." How could they boast, he asks, compared to the "vast surface of those unfathomable depths where man's mightiest works are but an atom"?[3] Despite their criticism, the lavish affair sparkled on in Newcastle, with balls, dinners, and 10,000 yards of red and white fabric wound into decorations, a gleaming pinnacle of decadence in a burgeoning factory town.

But 1838 also marked the dawn of the Chartist movement, comprised of political reform activists from the working class who wanted suffrage and an end to corruption. The years before and after it saw the advance of textile mills (from which all that calico fabric had come), and a labor force of children under the age of fourteen, which made up to 57% of workers in some cases. A corresponding "coal boom" hit the north, too, as demand for industrial and personal use rose precipitously. Newcastle's River Tyne offered the chief English port for its exportation, and great floating piles of black rock drifted down on their way to London. Newcastle was also home to the Heaton mine where flooding led to the loss of seventy-five lives, thirty-four of them young boys. They had not drowned, but were instead trapped alive to suffocate slowly on foul air in their living tomb.[4] Over five thousand children worked the mines of Cornwall, a figure that continued to rise after 1838 because they offered cheap labor to factory owners, were "obedient, submissive, likely to respond to punishment" and—perhaps above all—"unlikely to form unions."[5] Small nimble fingers and tiny frames could scamper in tunnels and avoid the slicing edge of dangerous machines: children were a commodity. Well into the 1880s, the years of Tesla's shining success, more than fifty thousand boys and eighty thousand girls would be employed in Britain's textile and dye factories, what poet William Blake called the "dark Satanic mills." Thomas Carlyle's *Sartor Resartus* didn't blame the industrialization by itself; he placed fault with the very engineer-scientist, whose "progress" worked to "destroy Wonder."[6] From the lowest beggar to the day laborer, to the clerks and professionals of a rising middle class, real human beings had to live with the changes wrought by genius minds. In this new world, "the living artisan is driven from his workshop, to make room for a speedier, inanimate one."[7] It is wise to remember that for all Armstrong's generosity to his beloved Northumberland, he was still on the wrong side of workers' rights. His friendship with Armorer Donkin, member of the coal owners association and embroiled in battles with miners over child labor, long hours, and low pay, and his own refusal to raise wages or reduce hours at his Elswick works mar his reputation. Technology brought with it dread and chaos, proof that the same dark twin that arose with the first scientific revolution had not been idle. A storm was coming, and in the 1860s and '70s, it would erupt into one of the most significant industrial battles of nineteenth-century Britain.[8]

The first power loom appeared in 1785, designed by Edmund Cartwright, but it would be nearly fifty years before the machine became truly automatic—meaning the warp, the dressing frame, the lathe, and the shuttle worked by cams and gears. Imagine the endless clacking of a Volkswagen-sized contraption, the monotonous drone of its engine and noise, powered by steam and fueled by coal. Charles Dickens's haunting description in *Hard Times* of a melancholy elephant bobbing its head in chained madness comes close, but ink and paper can scarcely convey the fumes, the rumbling, the *noise*. The textile industry built major cities like Manchester and, through automatic looms running on steam power, mills launched the industrial revolution. But they did so at terrible cost. Before the power shuttles, guilds of highly skilled weavers produced the fabric, the colors, the patterns. These same workers were promised that technological advances would ease their work, saving them time and effort. In the beginning, it did just that. But owners soon realized they could hire cheap, unskilled labor to run the machines (including children) instead. Having paid for the expense of a long apprenticeship, weavers found themselves suddenly without work, and it was this, rather than a particular hatred of machines, that drove the "Luddites" to break looms. Machines might not be the whole problem, but they had made it possible—first for mill owners to replace skilled labor with unskilled, and finally, to replace the worker altogether. We can imagine it, because we have seen it in our own age; a factory closes on three thousand auto workers; it opens again to employ three hundred, the rest being handled by efficient machines. Engineering changes everything.

While our tendency is to protect ourselves from change (as Elting Morison says), technology creates its own expectations. If you don't believe it, recall the chagrin of airplane passengers suddenly deprived of Wi-Fi—as if we had always been guaranteed this invisible connection to streaming data while hurtling through the space at thirty thousand feet. Once gained, it is very hard to go back to a techno "before," and so we ought to proceed with incredible care. French technology theorist Paul Virilio, known for his work on engineering accidents, reminds us that all technology invents its own destruction. "When they invented the railroad, what did they invent?" he asks: "An object that allowed you to go fast, which allowed you to progress—a vision a la Jules Verne [. . .] But at the same time they invented the railway catastrophe." It's easy to understand why the bright lights and heady pleasures of the 1838 Newcastle meeting did little to lift the spirits of men and women hard pressed under industry wheels. They wanted change too. But of a very different kind.

In 1842, just four years into Victoria's reign, a young man named Friedrich Engels would take a position in his father's Manchester cotton mill—and the world would never be the same. "Working Men!" Engels would write, "to you I dedicate a work, in which I have tried to lay [. . .] a faithful picture of your condition, of your suffering and struggles."[9] Rather than drink "champagne and port" among the upper and middle classes, and waste time "in fashionable talk and tiresome etiquette," Engels spent his leisure hours in the dark halls and back corners and crowded rooms of a population oppressed.[10] He concludes the opening address of his first publication, *The Conditions of the Working Class in Great Britain*, with a deeply troubling assertion. Despite all their words to the contrary, the rising middle classes (men like Armstrong, who were lifting themselves to great heights through the success of their factories and armaments) did not mean to improve the lives of all men. In fact, Engels claims, "the middle classes intend [. . .] to enrich themselves by your labor while they can sell its produce, and to abandon you to starvation as soon as they cannot make a profit by this indirect trade in human flesh."[11] Factory life was but slavery by a different name, the same sort of trade in bodies that the 1833 Slavery Abolition Act supposedly put to rest.* It was an ugly history, and one that Engels begins with the steam-powered loom that fueled Luddite backlash earlier in the century.

Before steam-weaving there was "hand work"; wives and daughters spun yarn, skilled weavers put it to loom. They did not grow rich, but they could put money away, sometimes enough to purchase a piece of land to cultivate and so feed and increase their families. Engels describes a pastoral paradise of youth spent in innocence and breathing fresh air. He compares it to the German countryside from which he comes, where no violent fluctuations held them in thrall—but where their minds were "dead." The industrial revolution offered something more, to begin with, Engels explains. In the preindustrial state, humans "were merely toiling machines," unaware of the "universal interests of mankind."[12] We've heard this language before. Judith Drake worried uneducated bodies would be "mere machines"; Descartes argued that the body (separate from the mind) was in *fact* a machine; Julian Offray de La Mettrie and the materialists suggested every bit of the human was only a clockwork creature, a thing of pistons and moving parts. But Engels speaks more like George Shattuck Morison, connecting machinery and mankind through the types of power they exert or control. "In every instance," Morison explained, "changes are characterized by some distinct physical device by which man could either use his own strength better than before, or added another force to his own power"—and it is only this power that allows for "intellectual improvement."[13] Men cease to be machines, so goes the argument, the moment they have machines to control. But when

---

* Notably, the Abolition Act didn't actually apply to territories in possession of the East India Company (an economic and political loophole). It would continue in Sri Lanka and Saint Helena until 1843, persisted illegally into the 1850s, and by some accounts continued in various forms in Australia well into the twentieth century.

Engels walked the factory floor of his father's mill, he did not see men lifted from toil to greater heights thanks to these new machines. Instead, "hand-workers" were driven from one position to another, in the same way that factory workers today struggle to cope as one by one plants are closed or sent overseas. T. W. White's 1844 short fiction *Six Days' Journey to the Moon* picks up the thread of these lost workers; it offers a world where "machines have in a measure taken the place of men and snatched bread from their mouths because they work so much cheaper and faster." The narrator asks what became of the thousand workers replaced by machines, and his host "entered a long dissertation to prove, that they were infinitely benefitted by the cheapness of everything occasioned by these labor-saving machines"— though "if they could get no work, or were deprived of adequate rewards, it was of little consequence to them that things were cheap, as they would have no money to purchase them."[14]

Manufacture, said Engels, centralizes property in the hands of the few. Meanwhile, the masses trudge on, through filthy streets "through the difficulty of human turmoil and the endless lines of vehicles" where people "sacrifice the best qualities of their human nature, to being to pass all the marvels of civilization."[15] Henry Mayhew described the lot of poor workers in *London Labour and the London Poor*, collecting stories of street urchins rendered destitute by the loss of parents to disease or accident: "Sometimes we hadn't no grub at all [. . .] our stomachs ached with hunger" said a sixteen-year-old boy with vacant eyes. "[Mother] used to be at work from six in the morning till ten o'clock at night, which was a long time for a child's belly to hold out [. . .] I was eight."[16] The slum houses stand stark and bleak on unpaved streets, glistening with brackish puddles of refuse. The "narrow courts and alleys" offer nothing of hope, only "tottering ruin" beyond description; "walls are crumbling, door posts and window-frames loose and broken [. . .] heaps of garbage and ashes lie in all directions, and the foul liquids emptied before the doors."[17] Manufacturing towns disappear during the work week, swallowed by coal smoke clouds that choke the main arteries. Here, there is nothing for anyone but despair. Engels reported on child labor, on the maltreatment of young workers, floggings for missteps, whippings for running away. Sixteen-hour days, six days a week, some mills employed night work, and even suffered workers to run machines up to thirty hours at a stretch. Crippled workers appealed to commissioners, boys and girls with curved spines and bowed legs, malnourished and broken. Engels quotes a Leeds doctor as exclaiming, "I never saw the peculiar bending of the lower ends of the thigh bones before I came to Leeds. At first I thought it was rachitis,* [but now] express the opinion that they are the consequences of overwork."[18] Physicians and surgeons spent time treating deformity, and also dismemberment, malnutrition, and lung complaints caused by the wretched atmosphere. The high rate of child mortality in the mills speaks for itself; dark and greased, the air peppered with dust and grit, no windows, no fresh water, no air.

I have, myself, worked in a factory. A plant for making car mats, the cavernous space shook with noise and smelled of burning rubber. Each line employed three people. One tossed chunks of rubber onto hot griddles that sizzled like waffle irons; a carpet-layer (in heavy gloves) pulled it free, still steaming and smoking and acrid. Precut carpet, frazzled blue and red depending on the vehicle, was pressed into the hot mess to bond the glue— then the mat would be flung onto a belt for the final step. I worked the end of the line as a cutter/counter, receiving in my turn melty stacks of stinking mat, ripping away the excess rubber, and clicking the counter for rights and lefts. Hundreds of them. Thousands of them. It was mind-numbing work; thoughts were impossible, just the one-two, one-two, cut and count repeated over like a mantra. I'd chosen it as a summer job; over a hundred degrees on the factory floor, sweat dripped past the safety goggles and rubber burned through the gloves. I coughed carpet fuzz, wondered if "carpet lung" were a real condition, and tried not to let the rubber collect in piles or cool in the wrong shape. It was miserable enough for me, but the lines stretched in either direction—and most employed permanent workers, mothers and fathers

---

\*    Also called "rickets," a condition caused by lack of vitamin D, phosphorus, or calcium.

providing for families and relying on this one thing to keep food on the table. I scarcely worked two months before the first layoff, however. And a few years later, the factory had gone, its derelict shape shedding bricks into a weedy parking lot. I'm under no illusion that my experience was, already, far better than the nineteenth-century mill worker—the factory had windows, ventilation, proper breaks, and reasonable hours. The Manchester mill, by contrast, was literally rancid. But "the worse situation," Engels wrote, "is that of workers who have to compete against a machine that is making its way."[19] Because a body that could not keep pace was a body unemployed; a body employed was a ghost already—a shell, a husk, a nothing. So much for the promise of machines.

## Seeds of Anarchy

Let us return to William Armstrong once more. He may have favored water power and longed for hydro energy (something he experimented with at his estate at Cragside), but his own works were nearly as dangerous as the mines that fueled them. In 1854, a fire in a Gateshead clothing factory superheated sulfur until "it came out in torrents, like streams of lava."[20] The sulfur caught fire and spread to Armstrong's warehouse, housing a Molotov cocktail of nitrate of soda, tallow, sulfur . . . and gunpowder. The blast felt like an earthquake; five hundred people were injured and fifty-three killed, with property damages exceeding one million pounds.[21] Industry might have seemed a boon to the expanding wealth and power of nations, but to workers in the grit, struggling for safer working conditions and better pay (and not getting them, usually), every single innovation came at cost. The Pemberton Mill of Massachusetts collapsed, killing many instantly and burying six hundred men, women, and children alive in the rubble. New owners ignored the building's capacity and shuttled more and more operating looms into the space. Iron supports groaned and heaved, and lacking the flexibility of steel, the girders crumpled. The debris caught fire during rescue operations, and the trapped burned alive, still crying out to those who could hear them and do nothing to save them. Death by disaster, dismemberment by machine, or replacement for mechanized labor seemed imminent for many, though the *New York Times* lamented, "Let the calamity be a warning," of the "unavoidable dangers of machinery and steam."[22] The Pemberton mill lost 145 workers, with another 166 maimed, but—significantly—jobs for over one thousand people went up in smoke.

The industrial age began as an age of steam, but it was also an age of soot, ash, and disaster. The worker, Engels believed, must work to escape these horrors, must work—just like Oliver in *Whitechapel Gods*—to overthrow the masters. Their own survival must come at the expense of the bourgeoisie set on exploiting them; the middle class, those men and women bobbing on Armstrong's lighted terrace or tucking their sketchbooks under arm to wander the scientific and engineering exhibition halls, had become the *enemy*. If the seventeenth and eighteenth centuries offered chaos and filth and death by disease, the nineteenth century had somehow failed to do better—not, at least, for the working man. And so chaos returns to haunt, but in a new guise, as *anarchy*. Engels had begun with the plight of workers. He would go on, in 1848, to coauthor *The Communist Manifesto* with Karl Marx.

In America, the word "communism," calls up images of Josef Stalin, the Cold War, and *Red Dawn*. But Marx and Engels's work has little to do with the manipulations that came after, and much more to do with helping the most disenfranchised and downtrodden members of society: the laboring poor. Engels praises the earlier Chartist movement, calling it the true separation of the proletariat (working class) from the bourgeoisie (capitalist wealthy class). But the future, for Engels, lay in the combination of Chartist with Socialist ideals, a creation of English Communism. Only then, he claimed, "will the working class be the true intellectual leader of England."[23] The revolution had begun right under the noses of mill owners: "it is too late for a peaceful solution."[24] For all the brilliant leaps of imagination, for all the fiery sparks of power, the New Epoch set the stage for class contest—and conflict—and one of the most fiery centered around the very Elswick works that Armstrong made famous.

In turning his attention to the manufacture of armaments, William Armstrong turned a successful business into an expansive and booming one. It stretched nearly a mile along the river Tyne and employed increasing numbers; the town had octupled in two decades and "houses and streets were thrown up in a constant attempt to keep up."[25] The crowded slum tenements described by Engels in big cities like London and Manchester began to creep into the once bucolic districts until scarcely a patch of ground remained. The promise of work brought men from everywhere, including from towns where machinery had already pushed them from more traditional workmen's labor. At first, all seemed well for Armstrong's works and reputation. But the narrative spun by the talented Engels (who has been compared both to Charles Dickens in his ability to describe dismal realities and to Thomas Carlyle in his fervor to change them) began, increasingly, to take hold of the northern laborers. They had not read Engels (it wasn't translated until 1887). Rather, the very conditions he described begin, at last, to foment real resistance among workers—spurred on, in part, by the blind greed of masters, mill and mine owners. What they feared most aligned with what the slave-masters feared: revolution, anarchy, riot, revolt. They'd fed the beast themselves, nursing it on expanded hours and reduced wages, grown fat on dead children and colliery fires. And not surprisingly, it's from those same dark tunnels that the trouble first emerges, clawing its way to the light with militant force.

Before the end of the 1860s, growing dissatisfaction with a 59-hour work week (which could blossom into 70) ignited tempers. In Elswick, where wages were fixed by a district trade agreement, those hours were frequently the only thing they had to bargain with. Spurred on by miner's strikes, the Nine-Hour Movement took hold.[26] The first petition appeared in Wearside, where engineering workers argued for reduced hours so that they could improve themselves through education, mirroring in its way the complaint of Engels. What good is machinery if it doesn't free men from drudgery to expand their minds? (In reality, it was also a way of arguing for overtime pay on the fixed system, but the point was a valid one.) The petition was dismissed, and the workers went on strike—and after less than a month, the employees had their nine hours. It had been too easy. Riding the wave of their success, workers formed the Nine Hours League in New-castle, electing John Burnett president, and sending a letter to all employers of Newcastle and Gateshead asking for reductions to the work week.[27] It seemed little enough; from fifty-nine to fifty-four hours a week—surely none would deny them?

On Saturday, May 6, 1871, a statement was issued by the then-*Sir* Wil-liam George Armstrong, declining, with unanimous support, the worker's petition. It was cold, impersonal, a direct reversal of the open and supportive dialogues Tyneside labor disputes had carried in the past. The resulting indignation jeopardized the one thing Armstrong had been known for: mutual respect.[28] Henrietta Heald makes certain to distance Armstrong from the direct conflict in her history of the strike. He had given over the management to a man named Andrew Noble, himself an "imperious, autocratic" and even "volcanic" employer, who treated the Elswick workers as though they were but boys meant to jump at his commands.[29] That he would be facing off to John Burnett of the League, an intelligent man with socialist leanings, would result in one of the most fiery conflicts in England. But Armstrong wasn't without blame; he made no move to replace Noble, who accomplished his work at a distance, and he would soon find himself face to face with an opponent he'd never encountered before. The mine workers may have begun the hard work in the 1860s, but their militancy had been easily overturned; crime, as Engels wrote, did not build alliances or create lasting movements. Burnett's Nine Hours League wasn't going first to arms; they were going first to the *papers.*

Joseph Cowen's *Newcastle Daily Chronicle* carried the stories, showing the demands as legitimate and the masters as inflexible. The masters, he wrote, feared that the men "have got political and now that want social and trade power"—in other words, they want to become their own masters, with a stake in the game.[30] The employers would not budge, however, and on the 27th of May, 7,500 men went on strike. Men at Armstrong's Elswick works dropped their machinery and walked away, and Armstrong, in a fit of anger or despair, made one of his greatest mistakes. He fired two hundred workers who had never signed up for the strike at all, making an example of them. And suddenly, it wasn't just the *Chronicle* that turned against the arms magnate. "Masters who reply cavalierly by lawyers' letters," wrote the *Pall Mall Gazette*, "and act as nearly as they can like despotic governments against revolutionary bodies can hardly expect [. . .] the sympathy of the public."[31] Armstrong lost money each day the strike continued; the machines fell silent, all the innovations of his armaments and big guns (for which he'd been praised a year or two earlier) stood at naught. But his heart was hardened and he would not meet terms. "We are now determined to break this strike by importing foreign labour," he wrote, and in due course, his emissaries went abroad to Europe in search of willing hands.[32] He closed the schools of Elswick works, using them to house his foreign transplants and refusing to open them except to loyal workers. The cities, already crowded, became insufferable. Men poured in to work the machinery, some of whom went, themselves, on strike, and requiring additional replacements. And then, on September 11, the clamor of dissent in the north suddenly erupted into the national papers, when the *Times* attacked Armstrong for his despotic part in the struggle. He returned fire, suggesting that no men ought to have the right to wrest power from their employers—and that "more educated and intelligent classes" ought to know better.[33]

He might have done just as well to call himself the oppressor of Engels's free men—and Burnett brought all his socialist intellect to bear, enumerating facts and accusing the magnate of suppressing the truth to serve his selfish business interests. Rallies broke out all over England in support of the Nine Hours League, and the foreign replacement workers voluntarily began to leave the country. The workmen would win their case by October, forcing the proud Armstrong to bow to their demands. Burnett, like Engels, saw the wider significance: "the working-man by himself is weak, but [. . .]

when combined with his fellows he is almost irresistible."[34] For the Victorian scientist-engineers, the battle lines were drawn between deprivation and industry, a fear of being left behind and a golden future of wealth and success on a national scale. But the lines had also been drawn between anarchy and control, between the specter of labor revolt and the enforced and terrible control of systems—of machinery that crushed men under the wheels, and men who had become machines.

## *The World that Wasn't*

*"His furious oration was remarkably similar to the whistles, groans, jangles, squeals, the thousand unpleasant noises which escape an active steam engine; the speaker's rapid delivery suggested a projectile hurtling at top speed [. . .] the grating phrases locked into one another like cogwheels."*
—Jules Verne, *Paris in the 20th Century*

Engels and Marx's *Communist Manifesto* traveled further and faster than Engels's own *Conditions of the Working Class.* The *Manifesto* appeared in 1848, just as a series of revolutions broke out in waves across continental Europe. In many respects, it was a workers' revolution, a populist movement that demanded rights for the proletariat. In the German states, students led a march for freedom of the press and of assembly that was ultimately put down by the aristocracy, while in Denmark a constitutional monarchy was established after farmers and liberals marched on Christiansborg Palace. The most impacting of them all, however, occurred in France. Driven by urban workers, it demanded *droit au travail* or the "right to work," workshops for the unemployed, and a voice in government. The king of France, a bourgeois monarch of elites, had been indifferent to the needs of the people—a despot in absentia, not active in suppressing them but oblivious to their cares. They wanted a voice, they wanted what the Chartist movement in England had accomplished in terms of the vote, and they wanted an elected parliament. As Armstrong would do, however, Philippe turned a deaf ear to all concerns. In 1846, harvests failed, workers lost jobs, and prices rose exponentially. No stranger to revolution, France once again teetered on the edge . . . and in February, a populist uprising overthrew King Philippe. Universal (*male*) suffrage became law in 1848, in time to elect Napoleon's nephew, Louis Napoleon Bonaparte III, as president of the Second Republic. The class struggle that Marx and Engels laid out so clearly seemed, at this revolutionary age, to be coming to full flower: "It is high time that Communists should openly, in the face of the whole world, publish their views" and so demonstrate that it is they who truly represent the will of the people.[35] It would be published in English, French, German, Italian, Flemish, and Dutch.

It was the bourgeoisie, Marx and Engels claimed, that gave rise to industry, to machines, to technology and all the trappings of factory air and coal dust that go with it. Through the mechanization of all things, they had "stripped of its halo every occupation hitherto honoured and looked up to with reverent awe."[36] Inventors, innovation itself, here appeared as a tool of repression, for the bourgeoisie "cannot exist without constantly revolutionizing the instruments of production."[37] And yet, we can attest even in modern times how hard those impulses are to resist. Almost as soon as Napoleon III takes over, the conservatives organize under the call for "order"—for control, for a means of taking the provisional and slipshod government in hand. Workshops were closed, and the classes took to the street, workmen against soldiers in the "June Days uprising." Marx took notice. The middle class had only *used* the working class to usurp the king. Napoleon III and his government shifted to the right, and in direct opposition to the original intent of suffrage and election, he established himself as emperor. And again, we know this story: campaign promises lead to dead ends, and the status quo returns. By 1851, Napoleon III had dissolved the National Assembly and proclaimed that liberty "never helped to found a lasting political edifice."[38] It's a story we should all do well to heed; remember Nietzsche's warning about those who claim to hunt monsters only to become them. Napoleon III established an anti-parliamentary constitution, unassailable by the pet electorate, and set his sights on industry.

Paris, for all its political and social strife, was nevertheless physically on the verge of becoming the city of light and wonder that charmed Tesla on his first arrival: brightly lit boulevards and an eager anticipation of all things technologically new. The Paris Exposition of 1864 hosted fifty thousand exhibitors, mostly from France and Britain, and *nine million* visitors. Napoleon III wanted to impress the world with his modernist vision. Displays included France's colonization of Africa and Polynesia, steam-powered vessels, and model homes meant for iron foundry workers at Le Creusot.[39] Marxist realities would not come to be; the machine would serve the bourgeoisie, for Le Creusot had been riven with strikes over wages and working conditions—and the houses were now only available to those who bowed to industry, to their *masters*. The working class had been duped; they had lost to a façade of promises, of state-commissioned hot-air balloons that didn't impress so much as leak and stink, weave and zag, and threaten to dump their passengers from on high . . . or the fiction of Napoleon's submarine *Le Plongeur* which never once proved seaworthy. The exposition tried to claim Paris as the world's center of industrial marvels. But not everyone would be so easily impressed. For Jules Verne, the power of science and industry enticed in its fabulous voyages, but it could also terrify. The steampunk aesthetic we've come to love in literature had very little of the human about it under Napoleon III, nor did it under King Philippe, or under Armstrong's factory rules. What would become of the writer, the artist, the free spirit, and human mind that Engels promises as the right of everyone? Verne's only dystopian novel, *Paris in the 20th Century*, looks into the future with a sense of overweening dread. His hero, Michel Dufrenoy, inherits a world of unfeeling machinery. "He is what they used to call an artist," one of the bankers remarks, "and what we would call a ninny."[40] To broker his assimilation into the collective, Dufrenoy "was kept under severe discipline [. . .] in order to break any impulses of independence or artistic instincts."[41]

Sensitive, poetic, a delicate soul, Dufrenoy finds himself in a world where the humanities have gone, all art and literature buried beneath a systematic education meant to train cogs in the machine of industry. Charles Babbage once feared the sciences would be forgotten, but in Verne's dystopia, only science and engineering remain. "If no one read any longer," wrote Verne, "at least everyone could read"; under the mechanical system of education "legions of employees" carried on, controlled by the government as little machines themselves.[42] The Director of Applied Science at the universal institution even speaks like a machine, the "whistles, groans, jangles, squeals" and "unpleasant noises" of engines and industry. The engine itself is praised for its utility and cheapness, its ability to light the streets, glare and bald, its means of rushing one group of passengers to the next, never stopping to wonder or admire, all impelled by the "demon of wealth" without either "mercy or relief."[43] This "modern" world rumbles with engines for war, "gas cabs," and urban sprawl; Dufrenoy's uncle, Monsieur Stanislas Boutardin, thinks in numbers and moves "with the least possible friction, like a piston in a perfectly reamed cylinder."[44] A director of finance, Boutardin sees no division between industry and recreation, between the mechanics of business and the mechanics of life. To drive the point home, Verne describes Boutardin's wife as the very symbol of the steam-powered age: "She was the locomotive and he the engineer; he kept her in good condition, oiled and polished her, and thus she had rolled forward for a good half century, with about as much sense and imagination as a Crompton Motor."[45]

This is no place for a young poet, Verne asserts, and the reader knows already that Dufrenoy, his dreams, and his romantic love interest are doomed to fall under the wheels. Bernstein and other critics remark that the story doesn't get much right about the future—and in some ways, they are quite right. Verne's dystopia isn't a picture of the future at all. It is, however, a very apt depiction of the century in which he lived, at least from the perspective of those being crushed beneath it. Verne doesn't extrapolate to predict the future; he's looking backward at Napoleon III's rise, and what it meant for his present. The world that "never was," he tells us, is even now before us. And just as Burnett won his suit against Armstrong by virtue of pen and ink, it would be the writers who fought for the standard of humanism before the coming world.

One of Napoleon III's chief and loudest critics was a political conservative-turned-radical who argued for universal suffrage, free education, and the abolition of the death penalty. He was also then—and remains today—the most famous novelist and poet in France, author of *Les Misérables* and *The Hunchback of Notre Dame*: Victor Hugo. He boldly defied the new regime, and declared Napoleon III to be a traitor to France, resulting in his own exile to Guernsey (then controlled by the British Empire). It's no accident, then, that Dufrenoy's first act in the novel is to seek out Hugo's works. The bookstore is cavernous, "imperial," and nearly palatial. Heartened by what he assumed to be a never-ending supply of literature, Dufrenoy requests Hugo's complete works, calling him the greatest poet of the nineteenth century. The clerks respond with blank consternation—a shrug, a sigh: "Never heard of him. You're sure that's the name?" Michel begs that they look also for Balzac, Musset, and Lamartine, but to no avail. "No doubt," the clerk explains after a fruitless search, "these authors were obscure in their own period, and their works haven't been reprinted."[46] What the shop stocks instead, by great crate-loads and in endless rounds of versions, are works that Tesla might have relished: *Abstract of Electrical Problems, Practical Treatise for the Lubrication of Driveshafts,* and *Meditations on Oxygen*. Verne's novel received criticism for being apolitical, for lacking the "big brother" of George Orwell or even a "little brother" in the form of the state.[47] It is true that technology has absorbed everything, even the governing structure, into a hive mind of society—but the "critic" speaks through his absence. In real life, the exiled Hugo continued to write and work and criticize the dictatorship. Redeemed in 1870, Hugo returned to his homeland a hero. In a future unknown to Verne, he would lay the foundation for the Association

Littéraire et Artistique Internationale, organized to protect the artist and his work, every bit the patron saint of Dufrenoy and entirely opposed to characters like Boutardin. Verne's political critique, and Hugo's too, was that the machinery of technology and of government alike will destroy humanity by first destroying literature and art. And this is still the complaint of the steampunk maker and artisan in our own time, that technological moguls have given us soulless devices devoid of personality. Their argument is not (like Carlyle's) against technology, but against its misuse—its abuse by a system that privileged the machine over the human being. And that was the English Luddite's argument in the first place.

It's no surprise that Karl Marx found a foothold with many when he published *Das Kapital* in 1867, and a revolutionary named Reclus translated it into French shortly thereafter. By the 1880s, a British Tory turned socialist had distilled it for easy uptake by workers, offering a version that "quickened their outrage and galvanized their activism" with "the promise of social revolution."[48] In 1881, a good two decades before George Shattuck Morison would publish *The New Epoch*, a different sort of "age" was being announced. H. M. Hyndman's article "Dawn of Revolutionary Epoch" had rattled the cage of a middle-class renegade: William Morris. Morris had also read *Underground Russia*, an assassins' creed, and embraced the idea of the industrial "terrorist" as "noble, irresistibly fascinating" and combining "two sublimates of human grandeur: the martyr and the hero."[49] It was Morris who noted the value of fiction (dystopian and utopian) as a mean of shaping ideology. After all, the sentiments and the sympathy were already provided, so long as the hero might be either pitied or emulated.

Morris didn't go to Paris, but to London, to stir up the men and women chained to machinery (and on occasion, hacked to bits by it). He took to the impoverished streets like an evangelist, espousing tenets that, says Butterworth, aligned with Reclus's *Ouvrier, prends la Machine!* (or "Workers, take the machine!"). The essay blisters with hatred for artifice and calls the subdued worker back to a greater age. "We shall go back to the work of the fields and regain our strength and gaiety, seek out the joy of life again, the impressions of nature that we have forgotten in the dark mills of the suburbs. That is how a free people will think," Reclus proclaimed, "It was the Alpine pastures, not the arquebus [a French gun], that gave the Swiss of the Middle Ages their freedom from kings and lords."[50] The same sentiment moves Michel Dufrenoy's only humanities-loving family member, Monsieur Huguenin—except that, in the "future" world of Verne's twentieth-century Paris, "there is no longer any such thing" as trees and fields and fresh air. And interestingly enough, the elder Frenchman blames a *nineteenth-century* London: "We envied London's atmosphere, and, by means of ten thousand factory chimneys, the manufacture of certain chemical products—of fertilizers, of coal smoke, of deleterious gases, and industrial miasmas—we have made ourselves an air which is quite the equal of the United Kingdom's."[51]

A collection of working-class autobiographies compiled by James R. Simmons gives the stark and startling details of nineteenth-century factory life that seem to prove Verne's point. An Aberdeen laborer complains of the "small" lives they lead: "I have seen the race become diminutive and small, I have myself had seven children, not one of which survived six weeks; my wife is an emaciated person, and she worked during her childhood, younger than myself, in a factory."[52] Another of the narratives describes the condition of the Litton Mill of Derbyshire, riven with contagious disease, where "excess of toil, or filth, and of hunger" led to fevers—sometimes as many as forty sick boys at once.[53] Likewise, the Lowdham Mill in Nottingham, with its long hours, bad provisions, and "want of wholesome air," led to precipitous declines in health.[54] Revolutionaries like Morris had landed in an echo chamber of woes, and though he believed that violent clashes might be avoided if the middle classes agreed to the tenets of social overhaul, he didn't expect anything of the kind.[55] Butterworth quotes letters from Morris, where he wrestled with these realities—"I want a real revolution," he claimed, "a real change in society [with] well-regulated forces used for bringing about a happy life for all."[56] His words resonated in 1884, where the newly formed Socialist League gave workers a hope for an ideal future. What followed it, we know too well—strikes turning violent, homemade bombs, true and terrible revolutions and wars and the communist empire that would rise from the ashes . . . but in Verne's imagined 1960, Huguenin's advice to Dufrenoy isn't to rise up. Rather, it's to "stay where we are, close our windows tight, and have our meal [. . .] as comfortably as we can."[57] The safe road had been recommended already, and by the emperor Napoleon III himself: bow to the mill, toe the line, and receive affordable, safe housing on display at the exposition. Verne's idol, Victor Hugo, had not taken that advice—and the young Dufrenoy doesn't take it kindly, either.

When Dufrenoy fails to learn the calculation machine, he is assigned as assistant to a man named Quinsonnas in "the Ledger." A mysterious place, Dufrenoy's new assignment so inflames his curiosity that he slips into the offices by night to see it. Groping his way ahead, he loses himself "in the center of this labyrinth" and brushes the Bond Safe in the dark. Suddenly, the floor gives way and he slides into the iron dark of a cage. An "apparatus of virginal sensitivity," the safe (through a series of trip wires and cogs) protected itself by "arresting" intrusion. Dufrenoy becomes a company laughingstock, but though the episode is comical and he escapes unharmed, the scene reverberates with violence: "a dreadful noise filled his ears; all the doors slammed shut; the bolts and locks slid into place" with "deafening whistles" and "garish light."[58] While never acting the part of a revolutionary, Dufrenoy symbolizes one. And in Verne's world, both fictional and actual symbols matter.

Anarchy gets its start in France with Pierre-Joseph Proudhon, who became a member of parliament after the French Revolution of 1848 (when Napoleon III was first elected president and before he suspended the constitution). Proudhon and Karl Marx corresponded, influencing each other's works—for both, the connections between labor, ownership, revolution, and anarchy were undeniable. The worker was *owed* the product, and should not be divorced from what they had made; Proudhon opposed ownership of property and other socialist-anarchists went further still, suggesting "the abolition of any sign of agricultural, individual, artistic or scientific property" and "the destruction of any individual holding of the products of work."[59] From these philosophies grew the "terrorists" admired by Morris, those who would sneak into factories and plant homemade bombs, or break looms, or incite mobs to the destruction of those industrial mechanisms that devalued the worker and consolidated property. Dufrenoy's night-walking raises the revolutionary specter, even as the lights and noise remind us of the disaster that befell William Armstrong's warehouse, the explosion and sirens that woke sleeping inhabitants of Gateshead, or to the Courrières mine disaster in France in which the workers were walled into burning shafts after rescue efforts failed. Henry George in the United States preached that, so long as capitalism allowed Vanderbilts to rise while the humble laborer lay down his hammer in dust, then "slavery was not dead."[60] Socialist anarchy received perhaps its greatest backlash in America after the Haymarket Riot, when five revolutionaries were sentenced to death after someone threw a bomb at police. One of the men, German immigrant August Spies, gave a chilling pronouncement before the floor fell out from beneath his noosed body: "There will be a time when our silence will be more powerful than the voices you strangle today."[61]

Dufrenoy seems unlikely to be a labor martyr; he is disposed to hope against odds, making friends with the pianist Quinsonnas, falling in love with young Lucy, and feeling that even the Ledger might be bearable. But Verne did not intend his novel to be one of acceptance in the face of mechanical and political tyranny, and Dufrenoy's first brush—soft as it may have been—against the devastation of machinery is only a taste of what awaits. The anarchists and Verne had this in common: mankind and machinery could not abide together peaceably. But for an increasing number of people, the anarchists were the real threat to peace. British psychologist Henry Maudsley attacked the French Revolution as the worst of mankind, and that too many freedoms actually caused more damage than careful strictures. (It was a common enough sentiment in Victorian Britain, a backlash of reserve and politesse in the wake of a rather bawdy and boisterous eighteenth century.) But there were tensions erupting in "alarmist" fiction, too, novels like Charles Fairfield's *The Socialist Revolution,* which blames immigrants and foreigners for inciting terror and degrading society. Long before the communist revolution, the laborer had taken on a suspicious aspect. The public, and the factory owners, had seen the violence of strikes and their repercussions. Labor reformers claimed workers were victims, exploited by capitalism and by the rich, but Verne's novel captures the insurmountable gap of sympathy—not only among the wealthy Boutardins, but even among Dufrenoy's closest friends.

Cast on the street for accidentally defacing the Ledger, and cast from the protection of his mechanical uncle, Dufrenoy believes he has tasted freedom. He tells his friends that his fondest desires are "life in the open air," "the love of a woman," "detachment from all ambition," and "the creation of a new form of beauty."[62] It's of a kind with the pastoral suggestions of revolutionaries, who recommend the laborer to the field, breaking from the politics of industry and gain. "Nothing happens by itself in this world," Quinsonnas explains, "as in mechanics, you must consider the milieu [. . .] a good engineer has to take everything into account."[63] Having invented a transforming piano/table/bed/commode, Quinsonnas's marriage of music to practicalities offers a middle road that Dufrenoy refuses. He will not sing for his bread, he tells them. But in aspiring to be "an artist in an age where art is dead," he falls victim to poverty and the shamefaced search for bare necessities. Dufrenoy even seeks manual labor, only to find (like the Luddites before him) that "everywhere machines were advantageously replacing human hands; there were no further resources."[64] Verne imagines the future well beyond the labor workers, the socialists and the strikes—he sees the triumph of Tesla's mechanic age as one entirely devoid of humanity. We shudder at the gutted sense of loss and shame: the young poet's "shirts yellowed and gradually fell to pieces, like leaves from the trees at the beginning of winter, and there was no spring to make them grow again. He grew ashamed of himself [. . .] he reeked of poverty [. . .] he would have inspired pity, if pity had not been banished from the age."[65] His story ends in the desperation of famine, as agriculture fails following a difficult winter. The penultimate chapter finds Dufrenoy clutching a single flower for his fiancée, and searching in vain for friends who have themselves been cast out. No one recognizes him, but he is followed anyway; he tries to run from the shadow of dread tech, feeling that "the demon of electricity" was pursuing him with sickening glare. He glimpses an electric concert, with electric pianos; he stumbles upon an electrocution of juvenile offenders by electric battery. It's a far cry from the wonder that Tesla expressed at Paris's gloriously lit boulevards; it's far more like the sickening descriptions of AC electrocution that rattled even Thomas Edison's constitution. And in the striking account of a frozen and starving France, Verne does his first real future-telling, unaware.

Three years after the Paris Exposition sought to put France on the map as a center for technology, Napoleon III's empire crumbled to dust. By 1870, rumblings of war with Prussia turned into outright support, with crowds in Paris crying, *"To Berlin! To Berlin!"*[66] Six weeks into the conflict, and the emperor faced his last stand, was captured, and exiled from France. In Paris, the laborers and revolutionaries cheered, and feminist journalist Juliette Adam proclaims "we shook off the empire as though it had been a nightmare!"[67] For a moment, the steampunk femme declaring victory for freedom and France tantalizes with what the future might hold; frozen there, we have the best example of those symbolic warriors who turn up trumpeting in fiction. But the heroine (and even the hero) in Verne's narrative is silenced by snow, freezing, emptiness. And following the apparent return of the Republic, the real Paris faces a cruel siege as well. Butterworth's history describes the harsh winter, closed off from resources. The "lucky" ones (including Victor Hugo, who had returned) ate the flesh of zoo animals and other exotics still within the city limits. Meanwhile, the populace grew "half starved and frozen, grief-stricken for infants who had died on a diet of cloudy water masquerading as milk."[68] Two days after Christmas, the German Krupp cannon (cousin to those displayed, ironically, at the Paris Exposition) rained shells on Paris in the worst of its bombardments. Those "machines of war" would lead to Paris's surrender, while revolutionaries desperate to get news out of France climbed aboard the unstable and stinking exhibition balloons to take their chances against the skyline and the pepper of gunshot. But this was not apocalypse. By the 1880s, Paris would welcome the likes of Tesla and Edison; by 1900 the Paris World's Fair (Exposition Universelle) would crown France's achievements with a medal featuring electricity towers, dynamos, photography, and the aeronautical zeppelin [Fig. 15]. Verne would live to see it too. And by then, he had become technology's greatest celebrant. Out of darkness, (artificial) light. For better, and for worse.

## Opposing Forces

As George Morison so aptly summarized, "It is hard to realize how rapidly the appearance of the whole earth may change." The old and the new cannot exist together. Or rather, the old must feed the new, must serve as the vital fluid of the very thing that replaces it. There are consequences. We ignore them at our peril.

In 1878, headlines proclaimed EDISON'S NEWEST MARVEL as cheap electricity. The light had come, but the coal must be gotten somewhere—and brought to the surface with as little expense as possible. "The essence of slavery," preached Henry George, "consists in taking from a man all the fruits of his labor except bare living, and of how many thousands [of men called] free is this the lot?" [69]

In 1881, Paris hosted the International Exposition of Electricity, featuring the dynamo engine. Meanwhile, laborers had been pushed to their limits (or pushed out of work) by those same machines. A few years later, socialist agitators Johann Most and Morris would hit the streets preaching the justice of dynamite and terrorist acts as anarchists paraded on millionaire's row. That same winter, the *Chicago Tribune* recommended that wealthy farmers poison leftover crops so the hungry scavengers couldn't steal them. [70] "A society corrupt to the core," Morris wrote, "engaged in suppressing freedom with just the same reckless brutality and blind ignorance as the Tsar [of Russia]." [71]

In 1884, Armstrong hosted a prince in his Cragside wonderland. Three years later, the same year Tesla sold his patents to Westinghouse, the Royal Mining, Engineering and Industrial Exhibition would celebrate Victoria's Golden Jubilee on same river that separated the common laborer from the scientific elite in 1838. A brilliant show, the exhibition occurred on the very heels of an April demonstration of nine thousand Northumberland miners for better wages, conditions, and rights—and just months before labor revolutionaries would be hanged in Chicago.

There are, we must concede, two histories—two stories, side by side and marching into the late Victorian future. One is shining wheels and bright brass gears. The other is coal-grimed faces, oily waters, and satanic mills. And at once, the steampunk aesthetic comes into steel-sharpened focus: *this* is the True Victorian Engine. This is the electric carnival with its dark underbelly, the bright tents hiding hellish weapons of war. This is Grandfather Clock's illusion of order, fueled in secret by the burning heart of Mama Engine. The Age of Machinery had come and there was no going back.

But there was still plenty of room for resistance.

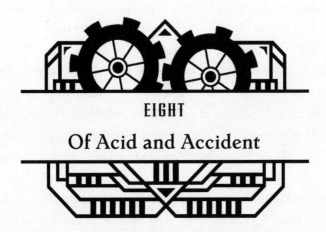

## EIGHT

## Of Acid and Accident

In the 1960s television series *The Wild Wild West*, steampunk hit the American frontier. A steam engine serves as home base to industrious Secret Service agents Jim West (Robert Conrad) and Artemus Gordon (Ross Martin), who fought criminals with high-tech gadgetry circa 1880. The 1999 movie adaptation went further still, pitting Will Smith and Kevin Kline against Western-Victorian villainy, complete with nitroglycerine-powered bicycles, flying machines, and a giant clockwork spider—but it's the television series that I remember best. And my favorite episode has far more to offer than just gizmos and gadgets.

"Night of the Bubbling Death" opens with swagger and suavity; two fine-looking gentlemen in their best Western-Victorian duds gallop into town on the Mexican border. The US Constitution has been stolen, and Agent

West leads an expert archivist to examine the villain's supposed booty. A revolutionary angling for independence from both Mexico and the United States, Victor Freemantle's hideout resembles a curious fortress—with a vat of boiling acid in place of a welcome mat. Declare the Panhandle Strip independent of its neighbors, and pay the ransom of one million dollars! West and Gordon aren't to be outdone; why return to Washington, DC, to negotiate when they can attempt a daring rescue on their own? The plot may be far-fetched, but it takes only a bit of imagination to recoil in horror as West dangles above the festering pool (from a wire cable conveniently tucked into his sleeve toolkit). The acid, a lurid orange-red, bursts in reforming bubbles. It may not have a mind of its own, but by comparison, the actual villain seems patently dull. Not very surprising, then, when Freemantle meets his doom in the treacherous ooze. Our heroes have more luck; they have more technology too. Sleeve guns, hidden explosives, smoke pellets—there are times when they seem as much Batman and Robin as Western do-gooders. Of course, Batman has his steampunk moments, as well. DC Comics' *Gotham by Gaslight* pits the Caped Crusader against Jack the Ripper in a Victorian Gotham City, reminding us that steampunk isn't confined to the British Isles. (Jules Verne's *Voyages* certainly weren't.) But there are other points of comparison. It wouldn't be Batman without an appropriate villain with a vat, and movie characters from James Bond to Hugh Jackman's reboot of *Van Helsing* feature grappling minions over bubbling cauldrons. Acid even makes its way into plenty of modern steampunk novels, too, such as Scott Westerfeld's *Leviathan*, and its "strafing hawks," birds of prey equipped with acid-filled spider silk to slice up airplanes. There comes a point when the discerning viewer has to wonder if every abandoned warehouse came automatically equipped with unguarded, skin-melting Jacuzzis. And the discerning viewer would be *right* (in a way). The hijinks of "Night of the Bubbling Death" might be pure fantasy, but that viscous vitriol over which West's feet precariously dangle is more than artistic license, and it comes down to something we might not think of as technologically significant today: *acid*.

When we talk about electricity, we have sparks in mind. But acid, as a technology, seeps into several different innovations at once, each with its own specialty. In 1865, Joseph Lister introduced the steam-propelled carbolic acid sprayer for use in surgery. Five years later, Lord Armstrong began building Cragside, the "house of the future" sustaining electric light with fifty-six acid-filled, power-generating tanks. But as with Free-mantle's devilish designs, acid had a far darker side. It could be weaponized. "Vitriol throwing" doesn't just appear in Arthur Conan Doyle's "The Illustrious Client," it eats away faces, appears at inquest, and zings through newspaper headlines—at the height of labor revolts, enraged and impoverished employees would turn to its viscous horror to attack job bosses and masters. Carbolic acid saved lives by killing germs; sulfuric acid (vitriol) destroyed lives by mangling faces. Acid, in all its forms, haunts our imagination whether it's burning through steel or powering the house of the future; on some level we know how effectively that same liquid would eat through flesh. Jim West escapes an acid bath; Victor Freemantle does not, but we don't need to see beneath the roiling liquid to know that brined bones are all that remain. Killer and cure, medical wonder and horrifying weapon: the rise of three acid technologies (and the tragedies, calamities, and accidents that follow) do more than inspire steampunk bravado—they prove that, though we may have two stories, they pivot on the same gears, rely on the same science. Every new technology invents its own perilous accident.[1]

## Carbolic Acid: The Technology of Cure

*"Frequently the acid was so poorly mixed that large globules floated in the solutions in which we bathed our hands or immersed*

*our instruments. The result was minor burns of the fingers and distinct numbing of the sense of touch."*
—Frank Emory Bunts, Marine Hospital, 1883

A technology need not have nuts and bolts and moving parts. Newton's calculus paved the way for the machines of the future, without a single brass dial or copper wire. And a peculiar property of phenol ($C_6H_5OH$) or *carbolic acid*, would lead to the single greatest advance in medicine before the twentieth century. And we can be glad of it, because surgery in the era of Armstrong, Edison, and Tesla was a blunt, perilous, and bloody affair.

Dr. Robert Liston may not have been the fastest gun in the west, but he was the "fastest knife in the West End."[2] In the 1840s, the best option for a gangrenous patient was the swift removal of the offending limb—and Liston had the record. Known for beginning a surgery by shouting "time me, gentlemen!" the good doctor enjoyed not only success but fame. His most (in)famous case, detailed by historian Richard Gordon in *Great Medical Disasters*, aptly demonstrates the principal problem of speed-racing a bone saw. The patient had been strapped down (a necessity, since writhing and screaming made surgery rather difficult) and Liston's assistant held on to the infected leg. The knife flashed, tendons popped, and the saw severed the limb in under 2.5 minutes; unfortunately, Liston had *also* severed the fingers of his assistant and slashed the coat of a spectator.[3] All three men died, two of infection, and the last from "fright." Richard Gordon suggests a heart attack—though we can't lay that strictly at Liston's feet. Still, according to Gordon, it remains the only operation in history to have a 300% mortality rate.[4] With odds like that, Liston sounds more like a practiced villain than a practicing surgeon, but his was an illustrious career. The accepted wisdom of the day ensured it: busted legs led to infections, infections led to death, and the quicker you could hack off the affected limbs the better your chances.

For me, the most shocking thing about Liston's checkered past isn't the mortality rates. Those were high to begin with—after all, the development of germ theory wouldn't get off the ground until experiments by Louis

Pasteur and Claude Bernard in the 1860s, some thirty or so years after Liston's famously deadly surgery. The idea that microorganisms lived on our bodies and could make us ill sounded more like demonic imaginings than hard science—like *vril*, or the pseudo cures of electric quacks. Liston himself published a case of "spontaneous mortification" in 1835, suggesting that the sudden shriveling and rotting away of an elderly woman's arm simply happened on its own: "You are well aware," he writes in the *Lancet*, "that at any period of life, and in every texture . . . death may take place to any extent."[5] That is, death just "happened," sometimes piecemeal, and most doctors assumed a stoic acceptance of mortality. No, the truly surprising thing about Liston's surgical career isn't that he made haste to remove limbs (or even that this occasionally went quite badly). It's that his *method* of speedy removal persisted even *after* the introduction of anesthesia to knock the patient out when, by all accounts, a surgeon could take his time. The reasons followed tradition and grim reality. Death by infection continued well after the introduction of germ theory, and doctors understood (correctly) that the longer they lingered in the body with their tools, the worse things were likely to go. Simply put: knowing that microorganisms existed, even knowing that they made people ill, didn't do the surgeon much good in practice. And that assumed the surgeon accepted the reality of germ theory in the first place.

People can prove surprisingly resistant to new ideas. Soap doesn't seem like a huge leap in technology, but though it was available in the operating room, many doctors simply weren't interested in scrubbing up. The ill-fated Ignaz Semmelweis, a Hungarian physician (1818–1865), discovered that pregnant women who delivered inside the hospital died of puerperal fever (or blood poisoning) far more often than those who gave birth in the filthy street. The mystery persisted until a doctor in the same institution died of fever, too, after cutting himself while dissecting a corpse. Semmelweis made the connection. As was the case in many hospitals, space shortages meant cadavers were located shockingly near the women and babies, and doctors frequently traded bone saws for forceps without pausing to wash. Something in the dead body had infected and killed the doctor; that *same* thing, he reasoned, must be killing women. While Semmelweis never gave these invisible killers a name, he instituted the shockingly unpopular policy of handwashing. It doesn't seem like much to ask, but soaps of the time were drying and harsh, made with sodium chloride, soda ash, or other agents. Doctors hated it, and Semmelweis's dictatorial insistence cost him two positions. Unemployed, he took to the streets admonishing women never to get pregnant. Seemingly unhinged, his behavior landed him in an asylum, where, in grim irony, he died of gangrenous blood poisoning (puerperal fever).

Of course, even for those committed to scrubbing away at fingers and instruments, many of which had wooden handles, soap alone didn't always do the trick. Two years after Semmelweis's death in 1865, Joseph Lister introduced the first link in the chain from acid to accident, but he did one thing more. Like the inventive minds behind Jim West's sleeve-gun, Lister also provided the means to dispense acid, making it practical for doctors to use.

The operating room of the nineteenth century wasn't much better than in the eighteenth century. No sterile sheets, no scrubs, and no rubber gloves. Surfaces consisted of wood, cloth, leather, and paper—all porous. By the end of the day, a surgeon's apron hung stiff with blood, and though instruments might be cleaned, they weren't sterilized between surgeries. In fact, a stiff apron was a sign of experience: surgery was grim work, pain a necessary reality. When Lister brought forth his ideas in the *British Medical Journal* in the 1860s, he faced timeworn traditions that went back to the sixteenth century and Andreas Vesalius. Many assumed that the air alone resulted in decomposition (or gangrene); essentially, infection was "brought about by the influence of the atmosphere." In the disastrous treatment of the American president James Garfield, Dr. Willard Bliss even praised "laudable pus" as a good sign (rather than evidence the president had gone septic). Pasteur had demonstrated the inverse, and Lister championed germ theory in paper after paper. It did not make him popular; doctors like Bliss remained suspicious even after Lister claimed to prove it: "The septic property of the atmosphere depended *not* on the oxygen, [. . .] but on minute organisms suspended in it [author emphasis]."[6] Oxygen merely fed the microorganisms and allowed them, in the right environment, to grow and multiply. The passive view of death "just happening" gave way to language that suggested virtual armies of germs were swarming healthy tissue. Lister's theory required a hostile intervention, a weapon. He chose a notably volatile organic compound to—as he put it—"exercise a peculiarly destructive influence upon low forms of life."[7] Corrosive, noxious, horrible to the nose and dangerous to the touch, carbolic acid proved a very effective weapon indeed.

Lister's earliest case studies applied the compound directly to the wound. It worked. It also burned away healthy tissue. He notes that even bone might be dissolved by the acid, but considering how often infection led to amputation and death, it seemed well worth the risk. Soon, Dr. *Lister* saved limbs that Dr. *Liston* would have hacked off. Even so, the acid cure could be, in itself, incredibly dangerous—if not to the patient, then to the doctor. Lister cautions that "carbolic acid and decomposing substances are alike," the cure and the killer germs both, induced *suppuration*, or a pus-filled abscess. In other words, an ounce of cure might very well be deadly.

## Misapplication and Mayhap

In February 1868, just a year after Lister published on the uses of carbolic acid as antiseptic, a woman in the "itch ward" of the Erdington workhouse was mistakenly rubbed down with carbolic acid rather than the usual sulfur and lime. She died in agony.[8] On October 26, 1870, a sergeant-instructor of volunteers named Patrick McGrath swallowed an ounce of carbolic acid after mistaking it for bitters. He could neither speak nor walk, and his lips, gums, and tongue turned ashy white. He struggled for air and lingered, suffering, for thirteen hours before death.[9] Clearly the antiseptic could be deadly when used incorrectly, but an article published in 1879 suggests that it might be just as devastating when used as directed. One Dr. Morrant Baker claimed that the method Lister recommended was "unnecessary and had serious drawbacks of its own." Others considered carbolic acid poisoning to be the "real evil," and several cases in August of 1881 seemed to prove it.[10] Two deaths occurred during ovariotomy, a surgery to remove the ovaries, and the *British Medical Journal* attributed the fatalities to use of carbolic acid (rather than surgical complications).[11] No one questioned the acid's power to corrode and kill; quite the opposite. But even those who saw its uses frequently misunderstood the purpose. Angus Mackintosh, one of Lister's students, claimed carbolic acid only worked because of the corrosion, and so any corrosive substance would do. As a chemical agent, he claimed, the acid encouraged the body to heal because it destroyed old tissue—a bit like encouraging growth by pruning a plant. Sound logic, but bad science; it would be similar to suggesting that chemotherapy works *not* because it kills cancer, but because killing cells somehow independently stimulates healing and is "good" for you. Mackintosh couldn't have been more wrong about the process, but he was right about the consequence. Carbolic acid's caustic quality did damage the fingers of doctors and assistants, scarring them, stealing sensation, and ending a few surgical careers—hardly a benign substance in the best of circumstances. Something must be done to mitigate its power, and so Lister devised an atomizer, a steam-propelled means of spraying acid into the atmosphere . . . simultaneously creating the first in a long line of acid "guns," still so prevalent in our steampunk psyche [Fig. 16].

The steampunk aesthetic is, above all, about craft. In the foreword to *Vintage Tomorrows* by Henry Jenkins, he describes steampunk fans as thinkers and tinkerers, Victorian-steeped "hackers" who wanted to give a material, corporeal "shape" to what they read about and imagined.[12] It shouldn't be surprising, then, to find so many items related to the genre (and to *cosplay* or costuming and outfitting the genre) on Etsy, one of the largest compendiums of handmade goods. Among the goggles and gizmos, corsets and top hats, you can find an assortment of steampunk "weaponry," such as Elisha's Acerbic Liquescence Atomizer. Andrew W. B. Elis developed this device for his Etsy shop TheGOAD (The Guild of Advantageous Development). Inspired by a computer role-playing game, Arcanum: Of Steamworks and Magick Obscura, Elis assembled the "gun" from a repurposed bug sprayer, clear glass jar, hand-wound copper tube, brass couplings, decorative front sight, and velvet-lined display box locked with a skeleton key.[13] The item is just for show (and would not fit up Jim West's sleeve in any case), but it bears striking resemblance to historical analog. And to be fair, Lister's carbolic acid sprayer, with its gleaming brass dome, glass carafe, and pinhole nozzles, looks far more like a weapon than a sterilization unit. Chief curator James Edmonson of the Dittrick Museum suggests that the sprayer probably worked like a kettle, heating and pressurizing water that, once vaporized, would turn the phenol into a fine mist. The wooden handle provided a means of positioning the device, which likely grew red-hot when in use. At roughly a foot and a half in height, the sprayer was portable but hardly convenient to carry around. They could weigh as much as 10 pounds, and the yellow cloud smelled of acrid tar.

Given the limitations, the carbolic acid sprayer could not distribute the mist with precision or accuracy. Unlike the Liquescence Atomizer, the device had no "site" mechanism and would never hit a target; it relied instead on overspray. Doctors' notes and letters describe "dripping" walls with carbolic acid clinging to surfaces like stinking beads of sweat. Frank Emory Bunts, one founder of the Cleveland Clinic, described the social consequences as well; having worked in a fog of yellow steam all week, his presence managed to clear church pews on Sunday. The smell lingered, becoming associated with hospital wards and making it impossible for doctors and patients to quite leave the operating room behind. By 1885, newer tech replaced the device, and many said "good riddance." *Aseptic* medicine had replaced *antiseptic* medicine; instead of spraying their hands, their beards, the wound, and everything else with foul-smelling vapors to kill the germs already there, physicians sterilized instruments and operating rooms, and provided gowns and gloves to keep germs out in the first place. The sprayer's tenure ended abruptly, relegated to the display cases of museums, but the idea—a weaponized means of spraying acid—carried on in fiction, as in fact.

## *Vitriol and Vengeance*

*"The vitriol was eating into it everywhere and dripping from the ears and the chin. One eye was already white and glazed. The other was red and inflamed. The features which I had admired a few minutes before were now [. . .] blurred, discoloured, inhuman, terrible."*

—Arthur Conan Doyle, "The Illustrious Client"

Arthur Conan Doyle's now famous hero, Sherlock Holmes, has appeared across multiple genres and—from Basil Rathbone's WWII-era Sherlock to the current BBC hit starring Benedict Cumberbatch as a twenty-first-century sleuth—in multiple time periods. Alan Moore features him (and his nemesis Moriarty) in *The League of Extraordinary Gentleman*, and he's even made an appearance on *Star Trek*. His predilection may be for solving crimes, but his career begins in chemistry. Test tubes and vials are hallmarks of the series, and in "The Case of the Illustrious Client," the doctor and detective face off with chemical weaponry. The key scene, one that Doyle manages with all the gusto of a medical man who has likely seen such cases, features a spurned lover's revenge on the inconstant Baron Adelbert Gruner. Her weapon of choice, sulfuric acid or, as it was known to the Victorians, *vitriol*, clings and burns, forever distorting the Baron's handsome features. The story exonerates the thrower (Kitty Winter) because Gruner's crimes against women and his intention to marry and ruin (or murder) a young lady with connections casts him as the villain. However, true tales of vitriol throwing rarely serve the ends of justice, and though the papers still branded it a "woman's crime of passion," vitriolic vengeance was far more common among the industrial contests—the "loom-breakers" and masters fighting over mechanization of factories.

Today, modern science recognizes *vitriol* as "hydrated sulfates" of iron, copper, magnesium, and zinc—meaning the crystals are mixed with water. Named *Vitreus* for the Latin word for glass, "oil of vitriol" (or "oil of glass") derived from heat-treating the salts.[14] A Sumerian word dating to 600 B.C.E. demonstrates its long history, showing up in discourses by Pliny the Elder (C.E. 23–79) and later by the Greco-Roman physician Galen.[15] For the vitriol thrower, however, it's the effects and not the history that most matter. "The crime of vitriol throwing," says surgeon P. Gaskell, author of *The Manufacturing Population of England* in 1833, "consists of putting into a wide necked bottle a quantity of sulphuric acid [. . .] and throwing this upon the person of the obnoxious individual." The caustic nature renders it "a formidable weapon," immediately destructive to skin and clothing.[16] *The Reformers' Gazette* carries a case study of one such attack, and the consequent execution of Hugh Kennedy in 1831. A servant in Glasgow, Kennedy "willfully and maliciously" threw a great quantity of vitriol on a fellow servant while he slept. The man awoke in agony, one of his eyes quite literally *burned away*. The gallows, says the editor, was the only thing the man deserved for the crime, which "has become so common in this part of the country, as to become almost a stain on the national character":

> It is so savage and cowardly, that fiends only, in the human form, can be guilty of committing it. The Spanish Stiletto is nothing to it; and we hesitate not to say, that if this crime of throwing vitriol, revolting as it is to the laws of God and man, be not henceforth and forever abandoned in this country, the severest punishment that one human being can inflict upon another ought to be applied to its guilty authors.[17]

The writer goes on to suggest what that might be, namely ripping their arms off so they can no longer throw acid. The outcry against vitriol throwing increased in proportion to cases, and cases coincided with industrial clashes. Near the century's end, George Shattuck Morison reflected on all that had come before and all that was likely to follow: "The danger is that the destructive changes will come too fast."[18] Men will lose their utility, argued Morison, they will not be able to adapt to new jobs as quickly, and the old jobs will be managed by mechanical means. Textile workers who saw their hard-spun labor being reproduced on massive scale by looms that needed neither to sleep nor eat—but that also had no hungry children to feed. Where the machine did not replace humans, it summarily enslaved them with economic shackles. In May Kendall's 1894 poem "The Sandblast Girl and the Acid Man" occupational hazards (including the dangers of acid glass-etching) mix with poverty to prevent successful family-making:

> I'm vastly better off than some!
>   I think how the many fare
> Who perish slowly, crushed and dumb,
>   For leisure, food and air.
> Tis hard, in Freedom's very van,
>   To live and die a luckless churl.
> Tis hard to be an acid man,
>   Without a sandblast girl! (ll. 65–72)

Verne's technocratic dystopia paints the same picture: the decades-long struggle between laborers and masters revolved not so much upon a right to wages as a right to live; this was war, and the battle raged on with vitriolic consequences.

Gaskell's work on artisans and machinery takes a long look at what he called "mechanical substitutes for human labor." The once thriving class, he suggests, had been reduced to "misery, demoralization, dependence, and discontent" in the wake of steam machinery.[19] Working for upwards of sixteen hours a day on about six shillings a week (about five pennies in today's market), they saw the advent of machines as an end to their livelihood, removing any chance for improvement. The hopelessness of it stretches from Dickens and Verne to the German expressionist film *Metropolis* in 1927: "It would have been well if steam and mechanism, in breaking up a healthy, contented, and moral body of labourers, had provided another body, possessing the same excellent qualities, as men," says

Gaskell—but the machines are hardly citizens, and instead the new means of manufacturing bred hostility and secrecy.[20] In the midst of this misery, the "spirit of revolt" rises, with murderous results. Though Gaskell's work essentially frees the masters (what he calls "their best benefactor, the enterprising and frugal capitalist") from blame, he also provides a Victorian view of what motivated and in some cases pardoned the vitriol throwers. The masses, who could not arm themselves with better, had found in acid the perfect malicious weapon.

If vitriol cases of the early nineteenth century primarily consisted of attacks against industrial magnates, why did Victorians consider it a "female" crime? Cesare Lombroso, an Italian criminologist (1835–1909), used elements of physiognomy, eugenics, social Darwinism, and a considerable dose of prejudice to suggest that criminals were *born* rather than made. "Criminal atavism," meant the degeneration of hereditary traits, a kind of genetic code for crime. Lombroso would assert that vice-ridden parents gave birth to vice-ridden children, easily picked out by virtue of their physical abnormalities (and thus privileging a specific race and class as having the most correct features and regular conduct). The implication? A hopeless inability to ever rise in society, and appalling treatment of minorities. But not everyone agreed. Gabriel Tarde shared Gaskell's (and Friedrich Engels's) views on the roots of poverty: environment has everything to do with it. But Tarde saw it as a crime of *imitation*. In his opinion, every evil crime originated in the urban center and then was passed (often through sensationalized media) to the rest of the region. And here is where the "vitriol as female crime" comes into circulation. Tarde blames Paris (considered by him a hotbed of vice, rather than Tesla's city of lights) for "the feminine idea of throwing vitriol in the face of a lover." The first instance, he claims, originated 1875 when a Parisian widow threw acid; the crime was shortly thereafter copied by others, and others, and others.[21] There are plenty of holes in the theory, not least because acid had been weaponized far earlier. Then again, copycatting itself made its way to newsprint in 1881. A young actress threw vitriol over her lover, and when asked how she thought of it, she claimed to have read it in the papers—in an article dealing, principally, with the outbreak of woman's revenge crimes.[22]

There were a number of cases where women attacked other women, sometimes perpetrated by unknown adversaries. In some cases, the effects were noted on clothing only after the attack, with no means of recompense of even of finding the perpetrator. Important exceptions—and this includes Kitty Winters from "Illustrious Client"—were those women who went after the men themselves, often for abandonment or infidelity. Historian Ruth Harris provides the details of several: a French tapestry worker named Amélia murdered her husband by vitriol for abandoning her while pregnant, a domestic servant attacked the father of her unborn child, and a woman by the name of Désirée attacked both her former lover and his wife.[23] At trial, most claimed to be in the right, justified by their maltreatment—but just as many were diagnosed as "hysterics," mentally deranged and dangerous. The media covered cases with apparent glee, and frequently memorialized the event with images. When the crime involved a woman, newspapers recounted the trial at length, savoring details of her charms, her clothing, her letters, her words, even to the point of sexualizing and fantasizing her character.[24] Vitriol throwing became synonymous with the exotic femme fatale. Though most throwers were industrial revolutionaries, targeting factory managers or the machinery itself, the press sensationalized the female crime.

It's an odd pairing: domestic deviant and loom-breaking factory worker. What would make acid their weapon in common? Ann-Louise Shapiro, in *Breaking the Codes: Female Criminality in Fin-de-Siècle Paris*, suggests the *vitrioleuses* were women "dangerous through their very domesticity—who transformed the ordinary and the womanly into the menacing."[25] This should sound familiar. Inanimate machines or automatons become horrific if sentient; what if they refuse their servile duties and swarm together to overthrow mankind? It's a common trope. *The Terminator, I, Robot, Metropolis*, and many, many others suggest that machines are the enemy. But in light of the disenfranchised masses clamoring at the gates for food, shut out by the (middle-upper-class) masters of automated factories, we have to question what *exactly* we fear. Is it the machine? Or the disenfranchised? Laborers outnumber foremen just as slaves outnumbered masters—and even women, should they come together for a single cause, might be just as dangerous to the established order. Vitriol, cheap and easy to obtain, became the weapon of the marginalized and subjugated, and in its ability to dissolve physical forms (the very mark of identity), it represented chaos and anarchy. It's Kitty's weapon against the Baron, and it's the weapon of the *literally* disenfranchised Victor Freemantle in "Night of the Bubbling Death" (he wants to be recognized as his own country, his own sovereign state). This is a world without edges, claims the acid thrower. In a gruesome and deadly manner, acid levels the playing field. Because acid is power.

## *Acid Batteries and the Power Principle*

The Stack Exchange's *Science Fiction and Fantasy* page asks: *Why So Little Steam in Steampunk?* So many of the gadgets appear powered by clock-work—and, in fact, "clockpunk" has been applied to the more da Vinci–like aesthetic. Even if we suspend disbelief, even if we allow for pseudoscientific possibilities, we can't quite get beyond the fact that steam power requires furnace, water, heat, and room. Steam power, like the carbolic acid sprayer, doesn't lend itself to mobile devices, but mobility and power frequently align. Jules Verne knew this all too well; you cannot have an extraordinary voyage if you can't get off the ground. *Around the World in Eighty Days* takes flight with fire-heated air, and *The Steam House* required fuel to power its steam-engine elephant into the thick of Indian jungle. But not every Vernian vehicle relied on Northumberland coal. In *20,000 Leagues Under the Sea* Captain Nemo powers the *Nautilus* with an unusually *vril*-like propellant. "There is a powerful agent," he explains to his captives, "obedient, rapid, facile, which can be put to any use and reigns supreme on board my ship. It does everything. It illuminates our ship, it warms us, it is the soul of our mechanical apparatus. This agent is—electricity."[26] We have tracked electricity's convoluted history from Galvani and Volta to Tesla. But though the *Nautilus* uses electricity for cooking, lighting, distilling water, and forward thrust, it requires no fossil fuel. *Undersea Warfare*, an official magazine of the US Navy, dedicated one issue to the ship's key features, from its conventional four-bladed propeller at the stern to its speed of 50 knots—to its remarkable source of power: a "hugely scaled-up elaboration" of a nineteenth-century *battery*, the "Bunsen cell."[27]

Invented in 1841 by Robert Bunsen (1811–1899), the battery uses a carbon cathode suspended in acid. Nitric acid and a zinc anode in dilute sulfuric *acid*, to be exact,[28] but Nemo makes modifications that turn a potential of 1.89 volts into something supremely powerful indeed: "Mixed with mercury, sodium forms an amalgam that takes the place of zinc in Bunsen batteries. The mercury is never consumed, only the sodium is used up, and the sea resupplies me with that. Moreover, I can tell you, sodium batteries are more powerful."[29] Sadly, some things only work in fiction, and the closest equivalent to Nemo's unlimited power is probably the nuclear submarine. But acid batteries were, in fact, the future. Without the means to store a charge, electricity would have remained a fascinating but unpredictable and short-lived event. With acid, power became portable.

Voltaic batteries, like the one that powered Davy's arc light, did not immediately find their utility in gadgets and gizmos (*Wild Wild West* notwithstanding). Tesla had begun to appreciate the possibilities through his Telautomatics, but the acid battery's earliest innovators turned away from the machine, and back to the body. Despite the constant vigilance of Faraday and the engineers, whose great aim was to make electricity respectable and who wanted nothing to do with plans for medical electricity, Galvani's "animal electricity" joined forced with Anton Mesmer's "animal magnetism" to turn the 1790s into a spectacular carnival of body experiments. Harry Lobb founded the Galvanic Hospital in 1861, and though the *Lancet* attacked Lobb as a quack, the *Electrician* hailed the place a roaring success: "an asylum for this science, a home from whence all our English discoveries will emanate."[30] His supporters came from the aristocracy, including Sir Charles Locock, 1st Baronet and obstetrician to Queen Victoria.[31] More importantly for the history of the medical battery, however, is Lobb's association with George Augustus Constantine Phipps, heir of the 1st Marquess of Normanby.[32] Lobb and Phipps corresponded at length over the "mysterious" condition of his daughter, Lady Constance Phipps. The lady, with her unknown malady, becomes a human test case for Lobb's outlandish ideas [Fig. 9]. Men might improve their virility and forgo exercise by having their muscles artificially contracted,[33] and women might be relieved of uterine diseases but also "nervous exhaustion."[34] Somehow, we've moved back to the electric principle of life, just in its more portable version.

Historian Iwan Rhys Morus describes Lobb's treatment as somewhat ambiguous, but it seems the good doctor had been using the battery to pass electrical currents through Constance's head and hand (where she was most afflicted with her mysterious illness). Unfortunately, we never hear from Constance herself, though her father writes that "she's made a wonderful improvement since you saw her" and that she might go down to London. In the meantime, "if you would like her to go on with the Galvanism," Lobb needed to send another battery.[35] The original had "broken." The process sounds, at first, like a step back, mimicking the experiments of Franklin and others in the eighteenth century (those who "drew sparks" the way doctors "drew blood"). But unlike them, this device could be used even when the doctor was not present and required no cord, no dynamo, no steam power to make it work. Lobb had medicalized the Pulvermacher chain.

Simple, small, and versatile, the chain consisted of a small metal plate and wire wound around a wooden dowel. The free end of the zinc wire is twisted into a loop while its other end pierced the dowel; at its opposite end, a loop of copper wire winds the dowel between the zinc and pieces the other side. The result, according to Lobb, was "sufficient tension, without the great heat and chemical power of large batteries"—in effect, a "perfect miniature" galvanic battery.[36] Lobb explains the utility in his 1859 treatise *On the Curative Treatment of Paralysis and Neuralgia*; when "excited" by acid, the battery generates electricity from positive to negative poles along a hydroelectric chain. For producing uninterrupted current, Pulvermacher's device could be set in motion by "clockwork" called the "electrophysiological modificator."[37] These strange chains of mini-batteries might be hung around the neck or stretched between parts of the body to administer minor electric charges, and Lobb was known to use chains as long as sixty links. His descriptions of the chain could be, with very little imagination, compared to descriptions of his lady patients, both "constant" and "easily excited," while the acid becomes the actor in an impassioned scene: it "runs up" to "excite" with "great energy," and is especially useful in the bath.[38] Men like Lobb continued to suggest electric batteries as curatives well into the nineteenth century, sold as chains or electrified belts, good for everything from headache to "female" trouble [Fig. 17]. Bizarre, brilliant, but essentially fairly useless as concerned Lady Constance, who (very probably misdiagnosed) died at the age of thirty-one. Perhaps ironically, then, German-English physician Julius Althaus makes the point in his own (far more popular) *A Treatise on Medical Electricity*: electric current "was not one of those remedies which, if they do no good, do no harm."[39] The trick with medical batteries is that, in effect, the body is a battery too—as well as a conductor: acid within the body, acid without. Lobb's popularity faded, and medical electricity would take a backseat to the progressive era of antiseptics and pharmaceuticals. But ambitious minds had bigger ideas for batteries than Pulvermacher chains; wizards like Tesla and Edison fought over power's manufacture. The wizard of Northumberland, by contrast, set his sights on the secondary problem: once made, how is power stored? We are back to the earliest question, but with renewed zeal: how to catch lightning in a bottle.

William Armstrong's northern estate, dubbed "the palace of the modern magician," may have included fire alarms, telephones, an elevator, Turkish baths and hundreds of *electric lights*, but in the mid-Victorian era, electricity remained inconsistent and unreliable, especially to a far-off country home like Cragside. How could he power an estate of such size without interruption? The answer would take him back to his early fascination with falling water, and would end in the complete alteration of the landscape. Charlie Pye-Smith describes the process by which Armstrong turned "1,000 acres of barren land into a magnificent, densely wooded, lake-strewn estate" and its rambling house. The lake was needed so there might be a dam, the dam so that there might be gravity-fed water, the water to power the dynamo or turbine—and all of it to create the first hydroelectric house.[40] Alan Morrison—former curator of Cragside Energy Centre—suggests that, while neither the discoverer of electricity nor the inventor of the water-driven turbine, Armstrong was the first to put the two together on such a scale. In 1878, Armstrong coupled a Vortex turbine to a Siemens dynamo in Debdon Lake—and two years later, Joseph Swan of Gateshead (Edison's chief competitor) installed incandescent lamps in the Cragside library, forty-five in all, though only nine could be lit at a time.[41] Dissatisfied with the problem of insufficient power, Armstrong took his electric quest a step further. He built a power house replete with a generator . . . and great, glass fish tanks of *battery acid*.

The tanks are, in themselves, worth the visit. They line the walls in banks of eerie blue-green, strangely luminous and reminiscent of fictional horrors. They aren't in use today (though the house is still powered by hydroelectric screw), but at the height of the Victorian age, each of the fifty-six viscous cubes served as lead-acid batteries, two volts each, to store power for uninterrupted use. Developed only a few years earlier in 1859 by French physicist Raymond-Louis-Gaston Planté and the chemist Fauré, these batteries could be *charged*. In 1859, Planté delivered a paper to the Academy of Sciences in France, claiming to have produced "a secondary battery of great power." It would also influence X-ray technology (just as Crookes's tubes had done). The battery used a sheet of lead and a sheet of positively charged lead oxide, separated by rubber. He rolled the whole into a tight cylinder and fastened electrodes to either end before submerging in sulfuric acid.[42] All batteries today are but advances on the original design, but there's no mistaking this for green energy. Noxious and poisonous, the copper rods grow crystalline growths of deadly coral.[43] Cables laid in wooden troughs took it from battery to house, a far cry from the wireless dreams of Tesla, more a representation of the same old tether that still holds us—linking lines between wants and needs, between the coal miner and the bright light. Armstrong never realized the dream of full reliance on hydropower; his estate burned gas to light less auspicious halls amid fantastic machines that failed, and fumbled, and needed to be serviced by human hands. (And meanwhile, the labor strikes continued, with their own version of acid power.)

Captain Nemo's dream of endless power and a symbiotic relationship between ship and sea evaporates in the reality of actual batteries—and their limitations. From dubious medical batteries to Armstrong's toxic fish tanks, acid provided power with consequences, the real bubbling vats behind the fiction. And the fact that such devices could, if so employed, render the human form unrecognizable was not lost upon the general public. While vitriol may have been the means of weaponizing the disenfranchised, some of the most notorious murderers of the nineteenth century turned to acid as a "closer" of sorts. The deed already done, acid became the means not of death, but of disposal.

## Dissolving a Murder

*Night of the Bubbling Death* offered TV viewers an image of how a caustic vat could get rid of pesky heroes and leave nothing behind. Dracula, in the steampunk-flavored *Van Helsing*, had his own steaming, acrid cauldrons, and (with the aid of CGI and an R rating) revealed floating, fleshless bones after a crew of evil minions went tumbling to their deaths. In the history of science, steam, and electricity, we've had plenty of heroes and rivals, but rarely do we encounter in real life the types of villains who would seek such a devilish means of dispatching their quarry. Not, that is, until the 1893 Chicago World's Fair. The fair—lit not by Edison, but by Westinghouse with Tesla's AC technology—represented the greatest achievements of the present and future, a suitable symbol of steampunk's greatest aspirations. Meanwhile, a villain of the first order was busy at the "manufacture of sorrow."[44]

Charismatic and attractive, the enigmatic H. H. Holmes, subject of Erik Larson's *Devil in the White City*, stepped off the train and into Chicago history. In a space of years, he tortured, murdered, dissected, and dissolved at least nine victims (though the count is suspected to be far higher) at his private "murder castle," the World's Fair Hotel. Fifty years later in England, the horror repeated when John George Haigh lured six women to their deaths in similarly grisly fashion. Unknown to one another, Holmes and Haigh nonetheless sought a similar answer to the principal problem of murder—how to get rid of the body. We began this chapter with Jim West dangling over certain doom, but the true history behind "Night of the Bubbling Death" is far stranger and far darker, and has roots in chemistry, medicine, and the very innovations the World's Fair meant to celebrate. Acid might be power, but acid also had the power to *consume*, to reduce a body to its constituent parts through grim chemistry. H. H. Holmes and Haigh (and others who copied their means to cover their tracks) relied upon acid for the grimmest of vanishing acts.

Chemistry and the chemical advantages of acid, generally, had been established long before the Victorian age, making up part of the tradition of alchemy and, further back, to the wonders of ancient Egyptian practitioners.[45] The field of *organic chemistry*, however, solidified in the mid-1800s, in large part to the efforts of Justus von Liebig and Friedrich Wöhler. Following Jöns Jacob Berzelius, most chemists believed that "organic" compounds (those formed from carbon, oxygen, hydrogen, and nitrogen) couldn't be created in labs—they were mysterious, they had a "vital force."[46] Von Liebig and Wöhler disproved the idea through their separate synthesis of molecules, and the discovery that arrangement of atoms mattered as much as number and composition. New combinations and the discovery of stable organic compounds earned Liebig his place as a founder of organic chemistry, his later work focused on agriculture—and upon food *production*. Agricultural chemistry studies production, or the processing of raw materials into food consumables. Liebig reduced the "building blocks" of plant life to its most basic nutrients (principally nitrogen), and as Clay Cansler points out in "Where's the Beef?," this spurred his interest in "more literal reductions."[47] Liebig spent the remainder of his career *boiling beef to jelly*. Turning cattle into potable, nutrient-rich, and easily accessible syrup for the starving masses turns out to be tricky business. The only thing more difficult than squeezing thirty pounds of beef into a pound of extract was actually raising the beef in the first place. The solution came from a partnership with George Christian Giebert and a cattle yard and factory in Buenos Aires (with a far easier climate for ranching than Germany). The processing consisted of "pulping" the meat, soaking and boiling it, and turning it into a thick, dark liquid that could be promoted not for use in the pantry (or not at first), but as *medicine*.[48]

Meat tea or meat juice may not sound remotely appetizing to many, but Liebig's company and the apothecaries that sold it recommended it for weak digestion and other ailments, a quick way to receive proper nutriment in liquid form. The Dittrick Museum has a share of ephemera advertisements and not a few physicians' tracts. Dr. J. C. Eno published *A Treatise on the Stomach and Its Trials* in 1865, suggesting "raw meat jelly" for the dyspeptic. "Valentine's Meat-Juice" offered a competitor's solution to Liebig's extract for the late Victorian in America, while in England, the Liebig name was "borrowed" by other manufacturers in their bid to provide the magic meat elixir. (Another of these mystery concoctions bears a strange relationship to Bulwer-Lytton and *vril*. Bovril may still be found on grocery store shelves in the UK. John Johnson developed the strange paste-like concoction in 1870 and took the suffix from the novel to boost sales.) But while advertisements hailed the extracts of meat juices as nutritious and beneficial for "wasting and debilitating disease," and while St. Thomas Hospital in London reportedly used over 12,000 jars of it a year, Liebig's extract ultimately proved a false hope.[49] By the time thirty pounds of meat had been scraped, pulped, boiled, skimmed, rolled, pressurized, and liquefied—very little actual nutrition remained. Historian Mark Finlay, in "Quackery and cookery: Justus von Liebig's extract of meat and the theory of nutrition in the Victorian age," even cites an experiment where dogs were fed nothing but the extract . . . And all perished of starvation and its complications.[50] As advanced nutrition, the liquefaction of animal matter failed. As a means of reducing bodies to easily disposed-of "meat juice," however, the process did have certain advantages.

## Hotel of Horrors: H. H. Holmes

H. H. Holmes arrived in Chicago in 1886. Larson re-creates the scene for *Devil in the White City,* describing his crisp suit, his good looks, his mesmerizing blue eyes (here meant quite literally). A physician names John Capen would describe Holmes in depth much later, using the physiognomy techniques popular at the time, and recalling his thin lips, dark mustache, and small ears—the overall effect might be compared to the one in Bram Stoker's *Dracula,* from the vampire's "hard mouth" and "heavy mustache" to his cruel and piercing eyes. In addition, Holmes "stood too close, stared too hard, touched too much and too long"; the effect, combined with his physique and manner, made him a favorite among the young women flocking to the city.[51] He set up shop as a pharmacist after "removing" the original proprietors (though never proven, most assume he murdered and disposed of the woman who owned it). He married in 1887, at least on paper; he had already married under a previous name, and he continued to court other women—including his bookkeeper's wife, Julia. Philandering was hardly the worst of his crimes, however. In 1892, the first act of a bizarre drama far more macabre (and indeed harder to believe) than any episode of *The Wild Wild West* unfolds: Mr. Holmes built the World's Fair Hotel.

Holmes himself served as architect. The hotel boasted one hundred rooms in a strange and sprawling structure that occupied a full city block.[52] The plans, pulled from the dark recesses of Holmes's murderous mind, defied logic (and not a few building codes). Janice Tremeear discusses its features in a book about "haunting" spaces: staircases that went nowhere, walls on hinges, blind passageways, rooms with no doors—or too many doors—or doors that opened to nothing.[53] The hotel also possessed hidden chambers, trap doors, and secret passages that no one besides Holmes could possibly understand. He had, in fact, ensured it by continually hiring and firing construction workers. When Holmes was captured, the dizzying tale scandalized and outraged the city of Chicago and, for a time, turned the World's Fair Hotel into a famed house of horror. He had murdered many of his own lovers, including the bookkeeper's wife, Julia, and her daughter, Pearl, and even his last lover, Minnie. How many others is difficult to imagine, but his cold dismissal of the crimes was perhaps most shocking: "I would have gotten rid of [Julia] anyway [. . .] I was tired of her."[54]

Holmes confessed to twenty-seven murders committed on site, though only nine were specifically proven. The reason has to do with remains—or rather, the lack of them. The disposal process varied. He practiced defleshment on some, meaning he stripped them of skin and muscle before selling the articulated bones to doctors. Larson's history links Holmes with a man named Chappell, who had mastered the art (and would do it for about thirty-six dollars); murder turned a relatively tidy profit that way.[55] Holmes dissolved other victims—turning them into the liquid flesh not so different from the "meat juice" elixirs popular at the time. The process itself bore certain similarities (macabrely ironic given that Holmes's hotel wasn't terribly far from Chicago's famous Stock Yards district). Liebig's elixir had been boiled and pressurized, though a small amount of hydrochloric acid aided the process (not that different from natural digestion).[56] Investigators discovered vats of acid with ribs and skull bits at the bottom, quicklime, a kiln, and a bloody dissection table with surgical tools.[57] Holmes had dissected the bodies, and then used heat or acid and lime to macerate the remains. The estimate of Holmes's crimes in those years during and after the World's Fair ranges from nine to over two hundred—but though acid aided disposal, it would be forty years after H. H. Holmes swung for his crimes in 1896 before the equally well-dressed and charming Haigh would earn the title "acid bath murderer," simultaneously returning sulfuric acid (vitriol) to center stage.

## Vampires and Vats: John George Haigh

How much acid does it take to dissolve a body? Typing this question into a search engine returns a surprising variety of (frequently very accurate) hits, from the television series *Breaking Bad* to mortuary science. The most recurrent entry, however, concerns a British serial killer of the 1940s. Described as suave and good-looking (rather like his predecessor), Haigh dissolved at least six bodies using a forty-five-gallon drum of sulfuric acid. The National Archives has the collected letters from Haigh, who was accused of murdering for monetary gain—but who, during his trial, claimed he was a "vampire" who killed in order to drink the victims' blood. In his final letter to his parents, he claims to be a martyr for his "religion," stating that "other nations are more enlightened than we are. They don't hang people for their religious convictions even though Sacrifice is involved."[58] His story about drinking blood could not be corroborated, and though he had hoped to escape murder under a plea for insanity, his claim was thrown out. A manipulative and compulsive liar, Haigh had previously spent time in prison for fraud, where he learned that sulfuric acid had curious effects—mainly its ability to dissolve animal tissue. He had experimented on rats using acid from the prison's tin shop; upon his release he set himself up—not in a hotel, but in a basement off Gloucester Road.[59]

In his biography of Haigh, Dr. Jonathan Oates remarks on the seeming glamour of the murderer. Attractive and well-spoken, and seemingly well-liked, Haigh swindled and cheated his way from one city to the next. To begin with, fraud seemed his primary mode of operation. Haigh delighted in the cleverness of his own scheming, and he steadily took greater and greater risks. Unlike H. H. Holmes, Haigh murdered men as well as women—Donald and Amy McSwan, and their son, William; Dr. Archibald Henderson and his wife, Rosalie; and Henrietta Durand-Deacon.[60] Also unlike Holmes, Haigh killed principally for money; he had no relations with his victims, though he did have a long-standing romantic friendship with a woman twenty years his junior named Barbara Stephens. Their love letters remain, and in them he appears courteous, gentle, and in every respect a kind man—and a loving son to his parents.[61] This, no less than his final confession of vampirism along with comparing himself to Jesus, Napoleon, and other "martyrs," caused journalist Stafford Somerfield of the *Daily Telegraph* to label him "one of the most baffling criminals of this or any other age."[62] Acid made his murders possible, not as a means of death but a means of escaping justice; he made a quarter of a million pounds selling off the properties of the McSwans (his abilities in forgery also handy in the crime) and various amounts for the sale of the Hendersons' personal items. In that sense, Haigh did "feed" off his victims, but with avarice rather than hunger, after money rather than blood. But acid, even of the sulfuric variety, doesn't dispose of everything. Haigh's final victim left a telling record of her demise, a trail to follow composed of three human gall stones and a dental plate.[63] Pathologist Keith Simpson—of the by then well-established field of forensic pathology—made a positive match, and Haigh was charged and brought before the Sussex/Lewes Assizes. The smooth, fresh-faced Haigh calmly explained that he should not be persecuted for his life as a blood-drinking vampire. Given so bizarre a claim, the papers originally dubbed him "the vampire," but this moniker didn't stick nearly so well as "acid bath murderer." A shame, in a sense, as he might have then been the "Sussex Vampire," completing our criminal circuit back to Sherlock Holmes's "Illustrious Client." In any event, both H. H. Holmes and Haigh proved the utility of the "bubbling death" for villainous intent. But we return to the Great Detective another way; Smith's work as a forensic pathologist sheds light on one final acid trick.

Deborah Blum's *The Poisoner's Handbook* details the "birth" of forensic medicine in the Jazz Age. But well before forensic chemist Alexander Gettler helped Charles Norris set up the field in the 1920s, toxicologists and other chemical experts were using acid to solve crimes—and to catch criminals. In 1836, the same year Humphry Davy began work on the "safety lamp" and the British Association meeting lauded Andrew Crosse, James Marsh had been called to assist in the Bodle Case. The whole Bodle family, it seemed, had been poisoned by tainted coffee. Arsenic poisoning mimicked cholera symptoms, which are devilish enough; once it enters the bloodstream, vessels become inflamed, and the victim suffers cramping and violent vomiting. The body tries to purge the poison, emptying the bowels, resulting in severe dehydration and burst capillaries. Since victims of poison and cholera died in similar fashion (and in similar agony), how could the casual observer ever tell the difference? Adding body tissue to a glass vessel with zinc and acid, Marsh incited a reaction for arsine gas. When ignited, the gas produced a telling silver-black metallic glaze, separating tragic but natural death from murder. It was the first test of its kind, but hardly the last; technology's progress is an arms race at best. And acid found its way into Gettler's lab, too, pulping pink organs into acidic solutions, allowing forensic specialists to sift through what remained. Dread tech, in the hands of detectives, becomes the means of tracing a mystery, a murderer, a villain.

If technology engineers its own accident, then acid provides for both ends of the spectrum: it operates as successful innovation and devious design, as germ-fighting wonder and vitriolic villainy. The steampunk tales in which acid appears span the same breadth; acid may threaten the heroes of *The Wild Wild West*, but it aids them almost as often (and eats through steel in the same episode to aid in their daring escape). Nemo's ship might be green energy, but the acid battery enables him to pursue his own deadly sort of revenge—and in Doyle's "The Illustrious Client," acid serves as weapon and justice in the same event. The evolution of this single innovation, appearing as it does in fiction as well as fact, paved the way for medical science, then for murderous intent, and then, once again, to a *particular* kind of medical specialist: forensic toxicologist. Anarchy and control find themselves wrapped up in the same technologies, the same drive for creating and maintaining power. And finally, a third figure, neither scientist nor engineer, appears on the Victorian stage. Forensic pathology rises from mid-Victorian laboratories to produce the most illustrious of steampunk of careers, the *detective*. In a "lofty chamber, lined and littered with countless bottles, [. . .] test-tubes, and little Bunsen lamps, with their blue flickering flames," Watson meets Sherlock for the first time, an absorbed student chemist in the midst of discovering a reagent "which is precipitated by hoemoglobin, and by nothing else." It would be thirteen years after the story's publication before such a test for "differential diagnosis" of human blood existed,[64] but Sherlock is ever ahead of his time. Like steampunk itself, the modern sleuth defies limitation, a retro-futurist subject standing between the old and new, a bridge figure in the age-old contest between our desire and our dread—and finally, between *death* and *immortality*. Many of our heroes, and a few of our villains, rise boldly into myth, appear in fiction, or (as with Tesla) compose a few fictions of their own. But in the face of death and misdeed, the greatest of Victorian heroes moves on a reverse track. Sherlock Holmes steps out of fiction and, seemingly, into fact.

# PART FIVE

DEATH

IMMORTALITY

"Of all ghosts, the ghosts of our old loves are the worst."
—Arthur Conan Doyle,
"The Adventure of the Gloria Scott,"
*Memoirs of Sherlock Holmes*

## NINE

# The Science of Sherlock

ome contests never really end. At our earliest beginnings, when humans looked up at the vast, wheeling sea of stars, they sought a place of changeless permanence in the face of corrosion, decay, and death. The darkness lurked as nothingness, a tomb, a fearful place of disintegration and loss; the light offered firm edges, solid ground, life itself. We brought light, and fire, and almost inevitable destruction to distant lands and dark corners; we worked in dim labs to generate sparks; we developed engines and industry; we sought control. But at each step, the threat redoubled. George Shattuck Morison wanted a world with no night, gleaming with created power delivered on rails and wires and through the ether to save us from the one thing we continue to fear, but never to name. Chaos, darkness, privation, anarchy . . . all of these are but other names for *death*. We die.

And we know we die. And we seek immortality. Science never achieves it, but heroes *do*. One of fictions' most famous resurrections does not require Dr. Frankenstein or the halo of electric charge. Sherlock Holmes, himself immune to superstition, dedicated to the scientific method, and committed to fact over conjecture, nonetheless returns to the living after his first demise—and continues, immortal.

"The real Sherlock Holmes," writes one nineteenth-century specialist, "would be a forensic toxicologist." As the writer himself (John George Spenzer) had this very position, it rings with a kind of pride of profession. But detection had not, to begin with, been a particularly scientific affair. The desire to sort crimes and punishments goes back through all recorded history; one of the earliest works on the subject, *Washing Away of Wrongs*, appeared in 1248 in China. It offered some useful advice (such as how to tell the difference between a drowning victim and a strangling victim), but relied principally on wise magistrates and good guesswork. The Victorian detective of the Scotland Yard variety arrives very late, preceeded only slightly by the first professional police force. The "Bow Street Runners" originated in 1749, founded by Henry Fielding, who was, himself, a fiction author—and their early cases do read like novel material.

On March 30, 1767, a gang of highwaymen waylaid a coach. They killed one passenger, beat another, and robbed the remaining two before disappearing into the outskirts of London. Within hours, seven men of the Bow Street Magistrate's office fanned out to gather information, and two days later five men were arrested.[1] The Bow Street Runners didn't just catch the men; they aided conviction by providing evidence in court. One man was hanged and later dissected by eager surgeons; two were transported to British colonies and labor camps. One turned state's evidence against the very men who had captured him . . . and the press covered it all in high detail. No one called the men detectives; the word wouldn't be coined until the nineteenth century, but they were stable, professional, and swift, with salaries paid by the government to protect the people.[2] By then, Bow Street belonged not to the novelist, but to his blind brother, John—a man who would reform and, in many ways, create the police as we know it. Called the "Blind Beak," John claimed the ability to recognize over three thousand thieves by their voices alone, an auditory Sherlock, but it was his desire for a paid force of professionals that made him remarkable. Criminals, he noted, had professionalized. It was time the police did the same, but it would be 1829 before a true London police force came into being (with its headquarters, Scotland Yard). They were better paid and better housed, but not necessarily better equipped, but controversy and scandal tarnished the reputation of the official squad and its sometimes blunt or brutal tactics. In 1873, referencing the trial of Sir Roger Tichborne, editor Edward Vaughan Kenealy wrote, "Can there be a doubt that Scotland Yard in all its departments is utterly useless for the duties of Police and Detectives?"[3] The science of detection, before the 1880s, was anything but scientific—unless, that is, you turned from the policeman to the chemist, and from the engineer to the medical man. As crime historian E. J. Wagner writes in *The Science of Sherlock Holmes*, forensics as we think of it today belonged not to policemen but to doctors.

Medical practitioners had risen from the early (terrifying) days of butchery and bizarre remedies to astute and analytical observers of the body over the previous century. And, given their experiences in anatomy, they could not only tell what ailed you while you lived, they could learn how and why you died. Andreas Vesalius opened the first wide window into the body through anatomy, but until the work of René-Théophile-Hyacinthe Laennec, few had been able to "see" into the *living* body the same way. In the early nineteenth century, Laennec worked at the Necker hospital in France. Hospitals of days past were not places of healing; quite the reverse. When *Whitechapel Gods* describes the slimed tools of the physician, prying loose gears out of flesh and expecting no great recovery, it's not entirely fiction. Hospitals were poorhouses, grimy with cancer-eaten bodies, wracked by the coughs of tuberculosis patients, and essentially a warehouse of soon-to-be buried wretches who couldn't afford home care. It would be well after Lister introduced the carbolic acid sprayer that such places took a turn for the better, but Laennec wasn't displeased with his office. After all, he could easily follow a patient from symptom to death to dissection table with a critical eye—and he had a new technology to help him do it. In the winter of 1816–1817, Laennec had invented the stethoscope. The Dittrick Museum has a number of early models—the very first was little more than a tube. But with this brand new device, Laennec listened to the heart and lungs of his patients, heard the rattle and bang of diseased tissue, and (after their inevitable decline) opened their bodies to discover the root cause of disease. It was far from glamorous work; hours spent with sputum-producing, half-drowned patients, hours more with knife and treatise, tracking cause of death. Laennec never collared a criminal, but the diagnostic practice of physicians led the way to seeing the body as more than victim. Corpse became clue—and the dead, from the perspective of the pathologist, *spoke*. The trouble was sorting out just exactly what they were saying. That took more than anatomy of a body; it required tests for determining how that body had met its end. It meant microscopes and test tubes and magnifying glasses and goggles; it meant turning the lens of science upon the workings of the human form. In short, it meant *forensics*, a new kind of body tech emerging from the same weedy beginnings of the clockwork boy and mother machine. Some of forensics history clearly owes a debt to the leavings of acid baths, but a great deal more has to do with caustic substance of a different kind. If London crime gave rise to the modern police detective, then the forensic specialist was born of *poison*. And the poison expert would come to play a role not only at the intersection of detection and medicine—it operates as the principal clue in *A Study in Scarlet*, with its trail of untraceable capsules.

As Deborah Blum writes in *The Poisoner's Handbook*, until the nineteenth century, "few tools existed to detect a toxic substance in a corpse."[4] Poison lurked everywhere from dyes to rat poison, and it was easy to get at; a murderer might be a cook, a maid, a wife, a father, a child. No license required, no questions asked. In France, murder by poison became so prevalent that arsenic was dubbed "the inheritor's powder." Metallic poisons like arsenic prevailed because, until the 1800s, no one understood the form or structure of the elements. But that was about to change with the advent of the "Marsh test" for discovering arsenic in organ tissue. James Marsh radically altered the future of the profession, not by virtue of a Sherlock mind, but by long hours in the chemical lab. What Marsh had realized at great pains did not, as a rule, result in scintillating courtroom drama. In fact, Marsh's process was messy and painstaking, and the slightest miscalculation could result in false positives. In chemistry, as in all other fields, it isn't always the inventor who profits most by his designs, but rather those who come after—people with superior skill, better contacts, or just plain better timing—those who turn first sparks into dynamos worth using. Predating Mr. Holmes (and in fact, most of the other legal medicine experts), we have the "father of toxicology," a man who arrived on the scene as a brand new kind of alchemist: Mathieu Joseph Bonaventure Orfila.

Like Newton and his predecessors, Orfila would spend hours in the dim back rooms with vials, eerie substances, and fire. Unlike them, he wasn't looking for the elixir of life; he sought the means of *death*. Having studied in Valencia and Barcelona, the good doctor spent three years testing poisons on thousands of animals.[5] Most of them died horribly, but the result of his exploration was the first complete toxicology of its kind: *General System of Toxicology, or a Treatise on Poisons*. It appeared in 1813, not long before Shelley's *Frankenstein*. In its way, the treatise was its own kind of "modern Prometheus"—not a God stealing fire for humans, but a man revealing for the first time exactly what a poison could do, and how to recognize signs of its deadly work. The old means relied upon testing stomach contents, but in substances like arsenic, which resulted in horrible retching, nothing would remain to test. Instead, men like Marsh and Orfila turned to the organs themselves, the once-pink lobes tarnished with metallic elements acted like the pointing finger of justice—a truly "tell-tale" heart, or kidney, or pancreas. In 1818, Orfila wrote a book on detecting and treating poison (accidental and otherwise), and suddenly his work had more than just forensic application. His fame assured, Orfila became physician to Louis XVIII, the last French monarch to die while ruling. Two successors later, and Napoleon III would declare his rule and

Verne would write his dystopia. Orfila would live just long enough to see that empire's dawn, but his ambitions were not political so much as practical, and his fame carried the day not with courtiers, but in courtrooms. In 1840, France's elite gathered to hear a celebrity case, as famous for its histrionics as for its dastardly plot: the murder trial of the heiress Marie-Fortuneé Lafarge. Orfila had his chance at public proof before an audience that included all of Europe.

The delicate beauty, Marie grew up a wealthy and prominent aristocrat. In *Forensics*, Val McDermid describes the twenty-three-year-old as young, possessing 100,000 francs in dowry, and eager to make a match.[6] It should have been an easy task. But there were as many adventurers in the marriage game as in the engineering sciences, and Marie's first taker, Charles Lafarge, was an inventor on the rocks. The technological advances across Europe needed more than coal; they needed steel. Pig iron could be produced by heating iron, coke, and limestone, but the resulting metal contains 3.8% carbon, while the more useful steel had only 0.4%.[7] The only way to reduce the carbon content was to use stove-packing and intense heat in a process that might take six weeks, and it wouldn't be solved until mid-century by Henry Bessemer and Bessemer steel. But that didn't keep others from trying—even in failure. Lafarge inherited an estate in decline, a crumbling manor, dingy rooms, and tangled financial affairs. He'd also inherited a desire to invent and very little common sense, it seems; he set up a forge on the estate, intent on developing new techniques for pouring steel. In the end, only his money flowed, and in only one direction: out. The forge failed, and on the edge of financial collapse, he spied the advertisement of a marriage broker: lovely young lady with the fortune. In due course, Lafarge wrote a vastly incorrect assessment of his worth and made his offer of holy matrimony. Within four days of the original meeting, the two were married and on their way to their new home.[8] What could go wrong? A lot, as it happened.

The house, lined in discolored furnishings, crawled with rats. Here were no lovely promenades, no admiring neighbors, nothing but people and places Marie considered "shabby and ridiculously old-fashioned."[9] Marie locked herself in her bedroom and demanded to be released from the wedding vows. Charles Lafarge must either let her leave the marriage alive, or carry her body away dead. "I will take arsenic," she wrote "for I have it on me."[10] The ominous words would haunt the both of them, in the end. While away on a business trip, Charles received a letter of seeming kindness from his new and newly aggrieved wife. She was mending fences, and with the note came a piece of Christmas cake. Soon after, he found himself too ill to continue his travels, and fearing he'd contracted cholera (known for its symptoms of heavy vomiting, diarrhea, and dehydration), he came home again. Maria nursed him ever so carefully, bringing him all his meals in bed while his condition strangely worsened. Doctors came and went, and all the while, Marie lavished affection on the failing man. He died on January 30, but in the end, he wasn't quite as duped as he seemed. Charles had made a second will. Meanwhile, his maid had noticed something odd being added to the food by her new mistress, and she kept a sample of Christmas eggnog in a cupboard jar *just in case*.[11] Two days later, the composed widow found herself under arrest. The prosecutor requested the Marsh test, but it proved inconclusive; they could not find traces of the element. Maria threw herself upon the mercy of jury, a poor injured widow, whose love for her husband none could seriously deny, a woman who had been at his side throughout. She cried, and she fainted, and her lovely figure, languid with grief, appeared in illustrated newspapers as a picture of tragedy. Readers pored over pages as though scenes from a novel, and public opinion divided. Had the law wrongfully accused this flower of womanhood of a devilish crime? The prosecution had other ideas—and a copy of Mathieu Orfila's book. He told the judge that the body must be exhumed. The rotting corpse raised, men brought the vile organ meat in for testing, causing spectators to faint from the "fetid exhalations."[12] But as before, the test failed them. In a last desperate effort to find justice for Charles Lafarge, the prosecution sought the only weapon they had left: Mathieu Orfila himself.

Val McDermid describes the scene as worthy of any television drama at season climax: Orfila arrived by the express train and set to work "macerating" organs into a rancid sludge of liver and brain. He set up the test as before—along with a test for the soil to ensure no compounds from the earth could be blamed for a false positive. Then, before the courtroom attendees, he performed the complexities of the Marsh test. This time, however, the results were positive. The difference wasn't in process but in method; as with the surgeons practicing antiseptic medicine, where the carbolic acid sprayer did no good if the tools, hands, and wounds weren't also thoroughly cleaned, James Marsh's test could only be as accurate as the practitioner. How was the evidence collected? From where? When? How much? In what ways did you heat or measure? The calculations could be tricky and easy to tamper with (they still are). But Orfila was not a bumbling surgeon, or a guessing detective. His steady hands and unperturbed temper, steeled by years of testing in the face of rot and horror, made no mistakes. The devilish "inheritor's powder" struck again; Madame murdered her husband to regain her dowry and whatever was left of the fortune. Thanks to Orfila's patience, Marie Lafarge spent the next decade and more in prison for her crime, where she died of tuberculosis. "Orfila's performance," writes McDermid, "would come to be seen as a watershed moment in the fight against murder by poison—a vindication of forensic toxicology."[13] But it was also the beginning of a new type of detective, one that understood chemistry, toxins, and the body itself almost as well as a medical man. It should be no great surprise, then, that Arthur Conan Doyle based Sherlock Holmes, who, by his own admission, "dabbled with poisons a good deal," on a peculiarly gifted surgeon, Dr. Joseph Bell. And of course, Mr. Doyle was, himself, a doctor. The Adventures of Bell and Doyle* doesn't have quite the same ring, but because of them, and through them, we have the greatest detective who never lived.

---

\*     Though, in fact, there *is* such a title—*Mr. Doyle and Dr. Bell* by Howard Engel.

## *A Study in Sherlock*

We first meet Mr. Holmes in *A Study in Scarlet*. The curious story, only half of which takes place in nineteenth-century London, the rest being a flashback through storytelling once the supposed poisoner is caught, hinges upon two curious clues. First, the word *Rache*, and second, two capsules found next to the last body. "Rache" means revenge, but Holmes addresses his inquiry at the capsules, discovering one (and only one) contains poison. We learn that the murderer has acted in revenge against two devious men from Salt Lake Valley, USA—but the poison pill would turn up as often as poison pens in detective fictions to come. Sherlock's knowledge of vicious substances appears in written tales, like "The Illustrious Client" and "The Dying Detective," but also in *unwritten* ones, the stories so often referenced by the duo in tantalizing one-liners. Why so much focus on chemical concoction? For the same reason doctors played the largest role in medical jurisprudence: Sherlock was a *chemist*.

Watson takes pains to describe a young Sherlock upon their first meeting: a "student," bent over test tubes and finding in that methodical work a fascination akin to treasure-hunting. "He is a little queer in his ideas—an enthusiast in some branches of science," remarks the man who introduces them in *A Study in Scarlet*; "a first-class chemist" but with "desultory and eccentric" habits.[14] Holmes is no medical student, however. Despite his long hours in the dissecting lab, he lays out his interest to Watson immediately upon meeting him. "Why, man, it is the most practical medico-legal discovery for years. Don't you see that it gives us an infallible test for blood stains. Come over here now!"[15] He drags the reluctant Watson to his station and proceeds to repeat the demonstration with the excited flair we might expect of a showman rather than a chemist. (Orfila did much the same thing in the Lafarge courtroom drama.) Though in fact no test for hemoglobin existed at the time, Doyle invents one, and Holmes takes obvious pleasure in being its namesake: "Now we have the Sherlock Holmes's test, and there will no longer be any difficulty." Murderers might be easily brought to justice, he reasons, because forensic analysis will pin a telltale rust spot to the per-petrator or victim with ease. As Wagner explains, Sherlock acts for himself partly out of convenience; there "being no forensic medical specialist at the scene," Holmes "simply fills the role himself."[16] And yet, in the foregoing series that follows, Holmes isn't remembered for his chemistry or his medical jurisprudence. He's remembered for an ability that would be a hallmark not only of the newborn detective novel, but of steampunk, and of the madcap history of science and power as well: he reasons *in reverse*. "In solving a problem," Holmes explains at the end of Doyle's very first tale, "the grand thing is to be able to reason backwards." Because time seems always to go forward, hardly anyone has a reason to think the other direction—it's the very problem Elting Morison references in his look at men, machines, and his great uncle's engineering prophecies. A memory that works both ways lets us see cause and consequence in a vastly different light. Elting calls it "recalling the future," an act of imaginative remembering that gives a picture not only of where we've been, but also where we are going.[17] In that way, *A Study in Scarlet* owes at least a small debt to the Red Queen. "Let me see if I can make it clearer," Holmes says encouragingly. "Most people, if you describe a train of events to them, will tell you what the result would be. They can put those events together in their minds, and argue from them that something will come to pass. There are few people, however, who, if you told them a *result*, would be able to evolve from their own inner consciousness what the steps were which led up to that result."[18] In his extraordinary way, Sherlock Holmes is a time traveler; he pieces together scene and analysis, fact and impression, working from the dead body to the motive and murderer days, weeks, months before the event. And while this all seems like the magic of an author's pen, the toxicologist and forensic specialist is a time traveler, too. In mid-Victorian England, amid the many future-hunting engineers in pursuit of lofty dreams (and their often deadly consequences), we find four very different heroes. Each of them, in their separate way, turns from the grand designs of the New Epoch and instead focuses on the minutiae: Joseph Bell, Hans Gross, Edmond Locard, and Bernard Spilsbury. The first two would inspire Sherlock Holmes—and the others, ironically enough, would be inspired *by* him. Art, imitating life, to imitate art.

## The Doctor Tells a Tale

Into the medical theater entered a bluff and husky workman, skin chafed from the cold of a November morning in 1878. He stands before the fresh faces of medical students and an angular sharp-eyed doctor. The newcomer shifts uncomfortably as those gray eyes take him in, fingers, toes, head to the careworn boots. "Your back—it's your back," the doctor tut-tuts. "But carrying a heavy hod of bricks won't improve it."[19] The man stumbled as if struck; all medical diagnosis aside, how could the doctor know he was a bricklayer? Did he see him with those eagle eyes as he struggled to carry the bricks, or follow him home where he could find no comfort in rest or sleep? Dr. Joseph Bell awed his students and his patients by diagnosing on sight, by revealing what a man or woman did for a living, or even what they had for breakfast, all merely by observing detail. But observation comes from more than mere looking.

"In teaching the treatment of disease and accident," Dr. Bell explained in his work, "all careful teachers have first to show the student how to recognize accurately the case." The recognition depended not on sudden swoops of intellect, but "in great measure on the accurate and rapid appreciation of small points." The student must be taught to "observe" and to "interest himself" is the context that surrounded the patient, but history to nationality to occupation. Sherlock Holmes famous assertion that Watson "sees" but does not "observe" rings through every letter. When Dr. Bell presented a patient to eager students, he gave them an entire narrative. But though Bell served as expert witness and collaborator on several criminal cases, he nearly always suppressed any notice of his involvement. As a physician and a "gentleman," he feared such dealings in police matters might sully his reputation. Bell would remain firmly fixed to the medical profession, unwilling to be tied to the more plebian police detectives. But the reputation of those very detectives was about to change, due principally to an eager student in Bell's medical theater: the young Arthur Conan Doyle.

In *The Sign of Four*, Doyle brings Bell's words to life through his now famous creation. Sherlock Holmes explains that the specialist was not to begin in the lab but end there. Instead, he ought to rely upon experience of his surroundings: "Let him, on meeting a fellow-mortal, learn at a glance to distinguish the history of the man, and the trade or profession to which he belongs [. . .] it sharpens the faculties of observation, and teaches one where to look and what to look for. By a man's finger nails, by his coat-sleeve, by his boot, by his trouser knees, by the callosities of his forefinger and thumb, by his expression, by his shirt cuffs—by each of these things a man's calling is plainly revealed."[20] It's what awed Doyle about Bell: forensics couldn't be just the inside of a body. Neither could it be only the scene or environment of the crime. Detection relied on both at once—detail within and without, the knowledge of a medical chemist, the eyes of an eagle, the nose of bloodhound. Details, the minutiae of everyday life, fuel the art of forensics like coal to an engine. Crime scenes speak to the investigator just as a body speaks to the physician, and in this steampunk drama of master-minds, medical tech, and malicious intent, the detective reigns supreme.

## From Doctor to Detective

Forensics of one type or another may be as old as crime itself, but crime scene investigation (CSI) remains astonishingly young—one of the youngest, in fact, of our "box-fresh" sciences. Ian Burney and Neil Pemberton's history of CSI in Britain consider it just over a century old, meaning it took root about the same time Sherlock Holmes did, sometime in the late Victorian age. Crime scenes "had always been there," they remind us, but they didn't have *importance*—they were not "substantive" in themselves.[21] How, their study asks, did the crime scene come to be thought of as a distinct space, a place of hidden clues? It doesn't start with DNA; if anything, it ends there. Burney and Pemberton suggest that our reliance on supposedly irrefutable DNA evidence actually narrows the legitimacy of forensics to a laboratory and a single set of tools—the pathology and microscopes "squeezing" everything else out of the picture.[22] Not only is this sort of modern view of forensics wrongheaded (DNA evidence is a fraught and delicate world just as prone to misinterpretation and accident as anything else), it misses the point. Crime scenes, like crimes themselves, are layered, nuanced, dynamic, and regularly "escape [. . .] the constraints of the lab."[23] For Burney and Pemberton, the very complexity of how we, in the modern era, have come to regard forensics "serves as a reminder that the past is a messy place." That was George Shattuck Morison's point—possibly even Newton's point—the "back then, back there" always has a helter-skelter appearance, full of rotten beams and uncertain paths. The problem arises from the fact that the present is also a messy place, and the future is likely to be just as messy; as a result, easy, uncluttered solutions are almost always going to be *wrong*. The investigator needed a model. Better yet, he needed a *handbook*.

Hans Gross, as Austrian criminal jurist, magistrate, lawyer, professor of law, and the man responsible for establishing the Institute of Criminology in Graz, Switzerland, compiled years of work as a "professional crime fighter" into one of the most comprehensive investigative handbooks ever written.[24] *Criminal Investigation: A Practical Handbook* published first in 1893—some seven years after Doyle's *A Study in Scarlet*. As an investigating officer, Gross readily saw the failures of police attempts to read or reconstruct a scene of crime. They were far too ready to insert their own narrative, to miss corruption of crime scene data, or to corrupt the scene themselves. They also missed the evidence available right beneath their (clumsy, defacing) feet: trace evidence. Like Sherlock decrying Scotland Yard for walking all over crime scene evidence, Gross insisted that a new protocol must be followed. Ian Burney and Neil Pemberton describe the salient features: first, the investigators must somehow "suspend" the crime scene in time and space, undisturbed and frozen; second, they must systematically "excavate" all trace evidence; third, they must do all with "absolute integrity," meaning no game of telephone that distorts facts as they are passed from hand to hand.[25] Practical, yes, but by no means simple. The first English edition ran to over nine hundred pages culminating in a "selected list of authorities" of more than one thousand entries.[26] The complexity comes from the very minutiae the detection ought to be cataloguing—how could the investigator keep from erring, from inaccuracies and his own bias? Like Sherlock, he is never to think forward in the absence of evidence, not even to guess at motive. To arrive at "incorruptible, disinterested, and enduring testimony," the detective must not only be "heroic" and in possession of youth, energy, health, liveliness, and vigilance, he must also learn "to solve problems relating to every conceivable branch of human knowledge," which might include everything from accountancy to knowing why a boiler might explode.[27] In other words, this new type of detective must be scientist, chemist, and engineer all in one. He must be Sherlock Holmes—and though the handbook and Holmes emerge almost simultaneously, there have been suggestions that while Joseph Bell provided the personality for the erstwhile detective, Gross provided his famous "method."[28] As in every history, however, personality counts for quite a lot. Grossian analysis would win the day, but criminologists who followed him found a more useful salesman for their science. Edmond Locard, French police scientist and the "Sherlock" of Paris, instructed his students to "read over such stories as 'A Study in Scarlet' and 'The Sign of Four'" to understand the new type of detective methods.[29] The bumbling doctor's accidental creation, like the semi-fictional stories of ocean-bound explorers, fired the imagination of a new generation.

Locard founded the Lyon forensics laboratory in 1912, and competes with other greats for the title of forensics' founding father. His great contribution lies in his reconstruction of crimes not from corpses, not even from murder weapons, but from *dust*.[30] Locard directed the Laboratory of Technical Police of the Prefecture of the Rhône, and this laboratory became an international center for study and research worldwide. But he was Sherlock's biggest fan. Locard literally fashioned his own working life upon the fictional character. He studied ink and handwriting, because Holmes uses it in "The Norwood Builder." He studied ciphers because they appeared in "The Dancing Men," "Gloria Scott," and "Valley of Fear." And Locard wrote a monograph on tobacco ash, principally because Sherlock had done so first.[31] In his 1929 paper "The Analysis of Dust Traces," Locard wrote: "I hold that a police expert, or an examining magistrate, would not find it a waste of his time to read Doyle's novels. For, in the adventures of Sherlock Holmes, the detective is repeatedly asked to diagnose the origin of a speck of mud [. . . and] The presence of a spot on a shoe or pair of trousers immediately made known to Holmes the particular quarter of London from which his visitor had come, or the road he had traveled in the suburbs."[32] Locard goes on to remind the studious criminologist that a spot of clay and chalk would originate in Horsham; reddish mud in Wigmore Street. The air around us swirls with our own past selves, skin and cellular material, grit from a shoe, sap from a tree, the waste and detritus of everyday: what, Locard asked, might we learn from it? "Sherlock Holmes was the first to realize the importance of dust," Locard wrote. "I merely copied his methods."[33]

Val McDermid's *Forensics* covers the history of pathology from Alfred Swaine Taylor through to Locard and on to another "Sherlock," London's Bernard Spilsbury. Swaine wrote *A Manual of Medical Jurisprudence* in 1831, but he was skewered by the press after condemning an innocent man through a bungled arsenic test.[34] The *Lancet* and the *Times* took Swaine to task and dubbed his practice "the Beastly Science," far from auspicious beginnings.[35] The latter-day Spilsbury, at work after Sherlock made detectives famous, had far more luck. The papers frequently compared Spilsbury, with his top hat and charm, to Mr. Holmes: no mere policeman, but a toxicologist, a lab scientists, pathologist, detective, and crime scene specialist. Media used readers' familiarity with Holmes as a way of calming panic in the face of often horrible crimes. In the Crumbles case (the murder of a pregnant woman by her lover), Spilsbury appears with unemotional elegance at the site of the dismembered corpse, "Cigarette in mouth, he was scrutinizing the sheets, clothing, and other articles,[36] and projecting an air of "cool indifference."

The crime may not compare to the infamous Haigh and his acid baths in terms of body count, but it had all the same gory details. Patrick Mahon murdered his girlfriend Emily Kaye. But he also dismembered her, attempted to boil away evidence, and left a trail of blood-spattered clothing and fat-clotted saucepans at his cottage in the Crumbles, a stretch of coastline near Eastbourne. Officers found parcels containing bits of her body, breasts and pelvis in brown paper, internal organs in a biscuit tin, bits of muscle and skin in a hatbox.[37] Investigators used Hans Gross's methods, painstakingly removing items from the foul-smelling cottage, and Chief Inspector Savage called the affair "gruesome and sickening [. . .] appalling."[38] On questioning, Mahon claims the two had quarreled, and Kaye struck her head on a cauldron. This, too, had to be taken in for analysis—and the next day, Bernard Spilsbury arrived at the scene. It would be the turning point of his career. Already known for his Sherlockian ability to "single-handedly assemble" the story of murder from a mortuary corpse (and make it stick in court), Spilsbury attracted media attention. Burney and Pemberton remark that the crime scene itself was not his natural habitat, however; a forensic pathologist, Spilsbury preferred the lab and mortuary—now, largely due to the horrendous condition of the body and its scattershot appearance all over the murder scene, he found himself in a makeshift tent and under prying eyes. The press arrived on the scene even before he did, and they stayed throughout. And despite Gross's warning that the investigator should avoid interpreting data to fit a story, the journalists had no such compunction. By "dramatizing" the forensic expert, say Burney and Pemberton, newspapers "provided ample material for storytelling and fantasy" and along the way, they "blur[red] the boundary between journalistic fact and crime fiction."[39] They created a story, in other words—as linear and decipherable as Spilsbury's polished courtroom delivery, and having just as little to do with the messy, tangled, chaos of the actual crime scene. Doyle's work also gave readers the story after its close, tactfully written by Watson and read before the fireplace in the comfort of the drawing room, a space of comfort and the illusion of control "over the chaos and horrors" of everyday life.[40] Doyle based Sherlock's methods on the reality of Joseph Bell; up-and-coming detectives (and perhaps more importantly, journalists and fans) based their personae on the unreality of Sherlock Holmes. "Without Joseph Bell, there would be no Sherlock Holmes," writes Bell biographer Ely Liebow, but without Sherlock Holmes, "few indeed would ever have heard of Joseph Bell."[41] Medical jurisprudence, the very seed of forensics as we now know it, flourished at the intersection of fact and fiction, because the chaotic world threatened disintegration, murder, death . . . and it would take a hero to order it again. The "pioneering hero of CSI," say Burney and Pemberton, the celebrity pathologist, formidable, knowledgeable, and (we hope) always in the right. It's little wonder the media dubbed him "the real Sherlock Holmes." But Sherlock still came first. And then, as now, he just as often steps out of fiction to supplant and replace his creator, and to compete in the public's estimation with the "real-life" detectives.

So who is Sherlock Holmes? Why does he appear here, in the dark channel between men and machines, steampunk and science, death and immortality? James Cabot and Brian David Johnson's steampunk history calls him the "CSI of his day, obsessed with cutting edge technology."[42] But in fact, Sherlock was the very *dawn* of CSI. Like Marcellin Berthelot, he is a chemist. Like Tesla, he has a laboratory to try out his inventions. Like Lister, he relied on the microscope, like Mathieu Orfila, he understood the human body and its anatomy—and like Hans Gross, he understood the *trace*. Technology guides Sherlock, and not only in chemistry; he relied on the electric telegraph, on fingerprints, ballistics, and handwriting analysis; he knew psychology and facial recognition and popular culture. He *did not* know much about the stars above him, but of the earth and its devices, poisons, pistons, technologies, and men, Sherlock Holmes was master. He may not have built a difference engine, but only because he didn't need one to solve a crime; he didn't invent the *Nautilus*, but only because he had no reason to travel under the waves. Of the once-living inventors, only Tesla comes close to matching his popularity in books, stage, and screen—but not even the wizard of electricity managed to have an original US TV series (*Elementary*) and a BBC series (Steven Moffat and Mark Gatiss's *Sherlock*) plus two blockbuster movies (Robert Downey Jr.'s portrayal in *Sherlock Holmes* and *A Game of Shadows*) at the same time, each with an eager, even rabid, audience. It is more significant still that Mr. Holmes and his belea-guered Watson float between time periods with no trouble; Basil Rathbone's Sherlock even hopped over decades to have the hero thwarting Nazis and no one batted an eye—while *Elementary* and *Sherlock* reimagine the duo as modern figures with remarkably little difficulty. I said at the beginning that we did not invent steampunk and retrofit it to the history of science; Victorian science has always been steampunk. When Sherlock invents a test for hemoglobin in *A Study in Scarlet*, few of his readers would doubt that it had in fact been done; Sherlock speaks into the future, and so in looking back at him in context, we are always somehow meeting out of time. He never dies. He never goes out of fashion. And even when his active presence is nowhere to be found, he is there all the same.

## A Most Extraordinary Gentleman

*"I gripped him by the sleeve and felt the thin, sinewy arm beneath it. 'Well, you're not a spirit, anyhow,' said I. 'My dear chap, I am overjoyed to see you. Sit down and tell me how you came alive out of that dreadful chasm.'"*
—Arthur Conan Doyle, "Adventure of the Empty House"

In Alan Moore's graphic novel *The League of Extraordinary Gentlemen* (considered a steampunk masterpiece of art and story), a collection of fictional characters join forces as a sort of literary Justice League. Mina Murray, the heroine of Stoker's *Dracula* acts as the brains of this unit of "gentlemen," the tie that binds Jekyll/Hyde, the Invisible Man, adventurer Allan Quatermain, and Captain Nemo. Each of them has been plucked from the pages of Victorian fiction to serve a mysterious "M" on a mission for England, but Sherlock Holmes never officially appears in the story. Though Mycroft, Moriarty, Sebastian Moran, and Watson form the backbone of this tale, the great detective has already met his (supposed) fate at Reichenbach Falls. But his very absence has a presence. Moore's greatest magic trick isn't in bringing supernatural and pseudoscientific forces together as a team, it's in fooling the reader into thinking that the fictions have been imported into a "reality" that belongs, not to Arthur Conan Doyle (or Telsa, Edison, and Crookes), but to Sherlock Holmes. Meanwhile, the world in which Doyle lived was full of ghosts, fairies, mediums, levitation, and voices from beyond the grave. In the end, the author—the man behind the fiction—vanishes into the fearful dread of Victorian spirituality, while Sherlock Holmes stands aloof, scoffing at superstition and wielding science and logic against all the murderous intent awaiting in the unknown dark.

Despite creating the famously logical Sherlock, Doyle himself devoted over forty years to the study of *spiritualism*, séances, and after-death experience. "There has [. . .] been no time in the recorded history of the world when we do not find traces of preternatural interference," Doyle wrote in *The History of Spiritualism*, but "the rising sun of spiritual knowledge" would fall first upon "the greatest and highest human mind."[43] The origin of spiritualism, he claims, lies with Emanuel Swedenborg, a Swedish scientist and mystic. A follower of Newton and contemporary with Sir Joseph Banks and Hans Sloane, Swedenborg saw visions and published his nine-volume *Secrets of Heaven* between 1749 and 1756. Doyle goes on to give his own accounts of spirits, hauntings, prophetic vision, the afterlife, and even the existence of fairies. The Harry Ransom Center at the University of Texas keeps a collection of Doyle's "spirit photography," eerie images and transfers that supposedly capture ectoplasmic shadows of the dead; the center also houses Doyle's personal Ouija board.[44]

Let's return to the Victorian séance. A modern observer transported back in time (says Alison Winter) would see in the performance something both trivial and peculiar. There are no sparks here, none of the whirling phantasm that the electric magicians would employ by the 1890s. Instead, they would find a subject, seated or reclining, and a mesmerist waving his hands about his or her features in "magnetic passes." Most mesmerists were men, many "patients" women, and a vein of scandal palpated at the idea that one might control the other as a living puppet, an automaton. However, the real attraction had to do with something deeper, a hope that through mesmerism, the subject would attain a kind of heightened awareness: "A new sense would open to her shut eyes," and the patient might "claim to see events occurring in the future, inside the body, in distant lands, and even in the heavens."[45] In other words, the mesmerist's trance allowed the average patient to experience something Nikola Tesla experienced anyway, all the time. But the real power of the spiritualist movement isn't only in expanded senses. It's in the uncanny ability to reach across a wider gulf than even the one that separates past and present: the one that separates life and death.

The movement began in the early hours of April 1, 1848, in New York (rather than in London). Margaret and Kate Fox, the same mediums investigated at length by William Crookes, claimed to communicate with ghosts. Knocks, rendered alphabetic, reveal a tale of murder and woe, and of burial under the Fox home. The story appeared in the *New York Tribune*, and soon, the entire nation was alight with powers first described by Mr. Swedenborg. The light that first glanced so brilliantly on the "greatest mind" had trickled down to the masses. Within seven years of the Fox bombshell, there were two million spiritualists in the United States alone.[46] It is hard to conceive of the power of so many voices, so many claiming to have heard from loved ones or seen a shade, to feel the nearness of something thought forever lost. Mesmerism, we remember, made claims of healing living bodies, too, like the electric fluid or even the imagined *vril* power.

William Crookes thought it a branch of physics, ill understood and waiting to be illuminated by science—and he came close to convincing Tesla, the materialist, of the same. Crookes compared their separate innovative ideas. Could electricity purify water? Could it travel through solid a substance? Could they set up multiple wavelengths and still ensure secrecy between two wireless operators? Could they control the weather?[47] Then, twirling the famous mustache (long and waxed into Crookes hooks), he added, "and what of mental transmission through ether between human receivers?"—communication with the living, and even with the dead. Crookes described his own experience of clairvoyance and levitation, and Tesla's materiality shook under the strain of these new ideas. As his biographer describes it, "the Old World swarmed through his brain like a hive of bumblebees,"[48] and all the superstition and dread, all those "unholy monsters," returned to darken his door. It was also at this dark time that Tesla received news of his mother's failing health. It is here, at the edge of death, that spiritualism most appealed; it offered a chance to blunt the sting.

Death. Always, it seems to us the closing of a door—we can draw near it, we may put out fingers to the handle. But once through, there is no returning. The immortal dream of Victor Frankenstein always crawls away, monstrous; even Victor's imagination stumbles over rot and worms, the disintegrating face of his mother and later his newly betrothed. The mid and late Victorians gained industrial power, but they moved further and further from the comfortable rituals of religious experience. Charles Dickens's works, always imbued with high morals, nevertheless give us human decency without the church, and even ghosts of Christmas strangely divorced from the actual dogma of Anglican faith. Urbanites were adrift in a sea of machines and smog-choked alleys—pollution so thick that Watson complained, "it was [impossible] from our windows in Baker Street to see the loom of the opposite houses."* Children died of lung complaints as often as from poor working conditions, mine collapses, or factory accidents. Signs of

---

\*    "The Adventure of the Bruce-Partington Plans."

death, from mourning clothes to black armbands, mingled in streets—and a penchant for mourning jewelry and memento mori peppered the commutes with lingering relics of the dead. Facing his mother's death, Tesla suffered bouts of amnesia, and oblivion, and at last was carried away to a separate building in a swoon. He lay there, helpless, but still under the deep influence of all Crookes had told him. "During the whole night every fiber in my brain was strained in expectancy," Tesla wrote in his autobiography; then, in the early hours, he "saw a cloud carrying angelic figures of marvelous beauty, one of whom gazed upon me lovingly and gradually assumed the features of my mother. The appearance slowly floated across the room and vanished, and I was awakened by an indescribably sweet song of many voices."[49] Weak and enervated, Tesla prepared to accept the otherworldly—to throw his old worldview away and, like the young Frankenstein, turn to other visions that offered something both more grand and more terrifying than the material world. There must be some reason for it all, some way to organize the chaos of loss. It's this hope that motivates Doyle too. Could you get "some account of my boy Kingsley?" he used to ask of spiritualists; he lost his only son in 1919, and would give his remaining career not to Sherlock, but to any comfort séance and mesmerism could give.[50]

"I never risk hallucinations," Doyle explained, "my experiments have had six, eight, or ten witnesses."[51] We can read in these words a reference not only to *evidence*, but also to a kind of jury—the eyes of men and women who can testify to what they see. It's by these means he "proves" spiritualism, and yet, as Sherlock Holmes remarks in "Adventure of the Norwood Builder," evidence may be "a two-edged thing, and may possibly cut in a very different direction to that which [we] imagine." Doyle visited mediums on several continents, and like Crookes, he attempted to use scientific methods to judge them. He refused to heed the criticisms of those who "had no experience" of psychic phenomena: "There's an enormous difference, believe me, between believing a thing and knowing a thing, and talking about things that I've handled, that I've seen, that I've heard with my own ears."[52] The results "justified" him; he could "fill a room of [his] house" with the letters of thanks from those who received consolation from psychic connection to lost loved ones—who "heard the sound of a vanished voice and felt the touch of the vanished hand."[53] Was this so different from the piles of mail he received for Sherlock Holmes from those who misjudged him to be flesh and blood? Weren't they writing letters to a ghost?

We might wonder how a medical man trained by Joseph Bell in observation of minutiae could believe such stories. Not all physicians were so credulous. Silas Weir Mitchell attacked it as both fraudulent and "very stupid"[54] (and later wrote a satire in which a man returning from war attends a séance of his legs and walks on invisible air, reunited). Tesla would ultimately turn his back on spiritualism, too, by discovering the external mechanism of his visions. The angelic faces, he wrote, were but the work of his eidetic memory, recalling a painting he had seen and admired; the music came from a nearby church choir at early mass; the likeness of his mother appeared only because she was so much concerned in his thoughts. These notes explained "everything satisfactorily in conformity with scientific facts," and he added, "I have ever has the faintest reason to change my views on psychical and spiritual phenomena, for which there is absolutely no foundation."[55] Regardless, séances boasted witnesses "of unquestionable intellect and social status," men like Crookes, but also Napoleon III and Tsar Alexander II (who were both patrons of the famed medium Daniel Dunglas Home).[56] Granted, neither Napoleon or Alexander received insight into the destruction of their separate empires, but even so, the spiritual had not risen as an *affront* to science so much as an aspect of it. Decades earlier no one would have believed electricity could light homes hundreds of miles from the source of energy—and for Doyle, his Sherlock and his spiritualism did not exist as if in separate spheres. In 1928, Sir Arthur Conan Doyle agreed to be interviewed on film. The grainy black and white can still be viewed today: Doyle, looking very much the part of Dr. Watson, sits on his front stoop with his dog, Dolly, and speaks about Sherlock and the realm of spirits. "There are a few things that people always want to ask me," he begins. "One of them is: how I ever came to write the Sherlock Holmes stories? And the other is about: How I came to have psychic experiences and to take so much interest in that question?"[57] He begins with neither. Instead, he described his scientific training in medical school—and his frustration with detective stories wherein lucky chance or "fluke" led to unexplainable results. "I began to think of turning to scientific methods," he explained; if Dr. Bell, his old professor, had instead worked as a detective, "he'd get the thing by building it up, scientifically." Doyle planted the seed of an entirely new kind of detective, and one so popular that he literally took on life of his own. It is "curious," Doyle tells his interviewer, William Fox, "how many people around the world are perfectly convinced that he is a living human being. I get letters addressed to him [. . .] I've even had ladies writing to say they'd be very glad to be his housekeeper."[58] Sherlock, he claimed, was a "monstrous growth" that even a skilled surgeon couldn't quite cut away—though Doyle had tried, principally by burying him at Reichenbach Falls. Ten years passed before Sherlock returned, years of angry letters from pleading fans, and even several attempts at copying the style and carrying the stories on without Doyle's help. He claimed that his more "serious" interest in spiritualism and the paranormal began at the same time as his work on detective fiction, about 1886. But by 1903, the fields would merge: he would bring his own hero back from the grave.

## The Immortal Detective

"It was in the spring of 1894," Watson begins, and London had been rocked by the death of Ronald Adair, "under most unusual and inexplicable circumstances."[59] But, he continues, "The crime was of interest in itself, but that interest was as nothing to me compared to the inconceivable sequel, which afforded me the greatest shock and surprise of any event in my adventurous life." Watson goes on to reveal his own continued interest in crimes of a curious nature, and as he mulled over the Adair case, he accidentally runs against an elderly and deformed man. A sneer is all he gets for his pains to help him gather the scattered books—but hours later, in his Kensington rooms, he finds the same white-haired creature at his door. In conversation, he points Watson's attention to the bookcase; a quarter of a moment later, and Sherlock Holmes stands before him instead. "A grey mist swirled before my eyes," Watson writes; he stood, he stared, he fainted dead away. When he comes to, his first response is to grasp Holmes by the arm to test that flesh and blood made up the man; we find Sherlock is no spirit at all, but rather a hunted man. The description that follows doesn't scintillate so much as disappoint; Watson accepts the whole, ready for another adventure (he is rather conveniently a widower at this point). But the reader feels strangely cheated. *That's it?* Yes, the tale ends with the famous capture of the last of Moriarty's men, the sharpshooter Colonel Sebastian Moran. Yes, we see Watson and Sherlock reinstalled at 221B. But if Doyle speaks truly that his hand was "forced" into pulling this particular spirit from the pit, then the somewhat unenthusiastic return may be evidence not of a grand séance but a continued haunting. All scientific method aside, the "Empty House" is haunted, after all. Sherlock outlives his creator. The Sherlock effect never lost momentum, even in his absence, and after all, *The Hound of the Baskervilles* appears in the interim between Sherlock's death and life.

"There's the scarlet thread of murder running through the color-less skein of life," Sherlock explains at the end of *A Study in Scarlet*. It is "our duty is to unravel it, and isolate it, and expose every inch of it." From his first case to his last, Sherlock offers concrete method, the "science" that Doyle felt was so lacking from detective fictions. From Edmond Locard in France, to Spilsbury in Britain, to John George Spenzer in Cleveland, Ohio, the leading forensic detective of the age offered advice from 221B Baker street. Tucked into Spenzer's archive is a typed sheet of paper, a copy of "Why the Modern Sherlock Holmes Would Be Scientifically Trained." The "new school" of scientific detective technique, it states, "May find Sir Arthur Conan Doyle a powerful ally." Every murderer of the modern age is confronted with a scientific problem—how to dispose of bodies. Many (like the acid bath murderer) had found in chemistry a solution, and woe betide to the detective who doesn't do likewise. "Good detective service is the product of an entirely different line of development [. . .] which involves a thorough knowledge of all the practical sciences."* The article cites the head of inspection at Scotland Yard, Charles Ainsworth Mitchell. Mitchell, advocate of handwriting analysis, fingerprinting, and new techniques, explains that Sherlock is but the starting point of a broad new field which, even by 1911, still had not been fully grasped. It's a case of innovation in need of a salesman. "Of all modern agencies," Mitchell explained in *Science and the Criminal*, "electricity is one of the most effective [. . .] for capturing the criminal."[60]

What, we might ask, does the manufacture of power have to do with detective work? Speaking like a latter-day Joseph Bell (and in agreement with Tesla), Mitchell explained: "The man in the street is not quick at grasping the possibilities of novel invention. At first, it is popularly regarded as a new toy, a matter of amazement and of amusement, but of no moment in the practical affairs of men."[61] He goes on to speak of all the inventions we've covered thus far in our journey, the telephone, the telegraph, electricity, and x-ray; every one of them was thought of no use, but "must prove itself and so live." And what proof mattered to men like Mitchell? "Not until the telegraph had shown its utility in capturing criminals did it acquire any reputation"; wireless transmission and even photography would have been but passing fads (in his estimation) if not for the light they brought to the darkest of matters: murder. The well-lit street discourages the criminal; the telegraph aids in transferring information so that he may be caught at railways by well-informed policemen; the photograph leaves no doubt that he is the man—and the microscope pins him to the crime scene in silent but effective witness. The future would see the rise of forensics not only in courtroom drama in Charles Norris and Alexander Gettler's** New York, but

---

\*    From the Spenzer archive, Dittrick Medical History Center and Museum.

\*\*   The principal characters of Deborah Blum's *The Poisoner's Handbook*, and famous for turning forensics into a systematic practice.

also in schools and police units across the globe. "If we presented Michael Faraday [. . .] with the scientific evidence our courts now take for granted," writes Val McDermid, "it would seem like magic."[62]

And perhaps there is magic in it. Sherlock Holmes, with his angular features pressed over fingertips, head haloed in pipe smoke, suggests that justice is possible, that truth may be found, that proof exists for those who know how to find it. The forensic detectives cannot bring the dead back to life, but they can make them speak, and through the dust and ash that surrounds us, offer a kind of order in the chaos . . . and at least the illusion of control. And of course, we can rest assured that Sherlock, who lived long before us, will be operating long after us, immortal as ever, the last "inventor" in an age of manufactured power (and one that George Shattuck Morison himself failed to predict). We long for the unemotional control of the Great Detective because he, alone among all the inventors, controls his technology. It can have no dread for him; he employs it, masters it as the Vryl-ya master the life force. But let us not forget: he—and the control he represents—is still always and ever *a fiction*. It's little wonder we prefer the fiction of an immortal Sherlock Holmes, steampunk hero debonair extraordinaire, to the mad science of our own dread tech, as much or more threatening now than ever it has been.

## FINISHING THOUGHTS
# Mad Science Reprise

What does it mean to stand at the edge of a New Epoch, surveying from that vantage the sweep of history long gone, and the vast blue dawn of a not-yet-imagined future? The end of Elting Morison's *Men, Machines, and Modern Times* offers proposals, but no answers. "Maybe in time," he writes, "there will be found some grand design to satisfy all the data in the disturbing new world, settled doctrine from which all men can take their bearings. But that time is hardly now."[1] In fact, as he explains it, the time is *specifically* not "now." The whole purpose for experiment in fact and in fiction persists in finding the potential and limit of machinery before it jumps the rails. We invent technology ahead of our capacity to contain or respond to new conditions; as a result, it will take "poets and philosophers and scholars," he says, to remind us of the strange bundle of contradictions we call *human*. Since we can't see far enough beyond the horizon of time, Elting asks that we "take the measure, a little more closely than heretofore, of what man is in the new environment he has created."[2] His great uncle never made any such promise, and now—some half-century later—we have more of George Morison's dash and daring than of Elting's caution.

At a dinner of steampunk authors, Diana Vick made the point that "authors can't write about the future because it's happening too quickly [. . . and] with steampunk we can go back and talk about things that didn't happen, couldn't happen, and there's so much more potential in that."[3] Then again, we can only imagine this potential backward. Like Sherlock, the story always happens in reverse. It's the speed of change that causes our dislocation—it's the train crash invented before the train. The New Epoch did come. It did bring many of the happy success Morison predicted. The trouble—the dread—has little to do with engineering prophecy. It has almost as little to do with the future.

There are those who would call the steampunk aesthetic escapist, a way out of the dingy cobbles and into the bright dazzle of something better, other. *Whitechapel Gods* may be escaping from the modern, but S. M. Peters's novel clanks along with machines of grit and destruction every bit as toothsome as modern-day weapons of destruction. Nemo might captain amazing vessels, but he's caught up in a very real political history of colonialism and imperialism we've inherited right to our present moment. In fact, the present never leaves us alone. William Gibson (of *The Difference Engine*) wrote *Pattern Recognition* in 2001, a year that splits history starkly between *before* and *after* the terror attacks of 9/11. The destruction wrought by four planes, aviation technologies that no one before considered as tantamount to missiles, returns us to Virilio's warning about the invention of accident. No hot-air balloon? No pitching into earth from above. No boiler? No explosion—no airplane, and New York's skyline might still have Twin Towers overlooking the harbor. "The attacks of September 11, 2001," write Rachel Bowser and Brian Croxall in *Like Clockwork: Steampunk Pasts, Presents, Futures*, "changed everything."[4] We've proven it a familiar story—World War II and the atom bomb, World War I and the catastrophic effects of gas and guns, Armstrong's "Big Will" or Tesla's radio waves, even the locomotive itself, which allowed borders to be crossed so readily and at such speed: you cannot reverse upon technology. We must live in the *after* with all the *aftermath*. But it's the way the past interrupts the present that Freud called "trauma." Theorist Cathy Caruth adds that "trauma is not locatable in the simple violent or original event," but that it comes back in unknown and unknowable ways to haunt us.[5] Steampunk is ever the present and past interrupting each other's narrative, and our dread doesn't rise from seeing some future danger, but living in a troubled present.

"Nobody really writes about the future" at all, said Gibson when asked about the book. "All we really have when we pretend to write about the future is the moment in which we are writing."[6] This helps explain steampunk's "moment"—its rise to national and international prominence. If speed of technological change dislocates us, steampunk beats it to the punch. Lacking the Red Queen's memory in fact, we invent it in fiction, recreating a reality where planes (or rather airplanes) *do* crash before we've invented them, but at least it no longer surprises us. Meanwhile, we go on inventing the means of destruction for a future generation to assimilate.

"Our generation," wrote George Shattuck Morison, "has the privilege of doing its full share in bringing forth the great changes which are ushering in the new epoch." The truth of our steampunk science lies just there: clockwork futures aren't about the future at all. They exist here, and now, the present moment with its darkness and daylight, its struggles and sorrows. The strange little book got a good bit of the present right. Smallpox eradicated, epidemics at such a low ebb in developed countries that people debate *whether* rather than *when* they will vaccinate. Earth movers and building techniques lift cities skyward with better ventilation and safer construction—and half of it accomplished by nonhuman enterprise. Children no longer labor in mines; prosthetic shops don't line our boulevards for the unlucky factory worker. We've seen better and cleaner means of living since the 1970s, used new materials to protect what's left of the environment, and proven that yes, both Armstrong and Tesla were right, we *can* harness the sun, and the wind, and the water for power. It's also meant pollution and war and damage and death. It always does. If steampunk science has taught us anything it's that we live precariously with our inventions and their intersections: golden gears and dreadful weapons, medicine that heals and poison that destroys, past and present, fact and fiction. It's best, perhaps, if we leave the haunting and—like the sleuth—take up trace hunting. We have paths to choose, here at the edge of our own epoch, the future ancestors of an inherited past.

If the science be mad, then let us be heroes.

# ENDNOTES

**PERAMBULATION: THE MAD, MAD WORLD**

1. Morison, George Shattuck. *The New Epoch and the University* (Oration, delivered before the Phi Beta Kappa Society in Sanders Theatre, Cambridge, Mass., June 25, 1896). (2)

2. Morison, George Shattuck. *The New Epoch.* (1896) Boston: Houghton, Mifflin and Co., 1903. (5)

3. Priest, Cherie. "Steampunk: What It Is, Why I Came to Like It, and Why I Think It'll Stick Around," *The Clockwork Century.* Aug. 8, 2009.

4. Jenkins, Henry. "Foreword." Eds. James H. Carrott and Brian David Johnson. *Vintage Tomorrows.* Sebastopol, Calif.: O'Reilly Media, 2013.

5. Whitson, Roger. *Steampunk and Nineteenth-Century Digital Humanities.* Abingdon, England: Routlege, 2017. (2)

6. Schafer, James, and Kate Franklin. "Why Steampunk (Still) Matters," (2011) Parliamentandwake.com/veil/whysteampunkmatters/whysteampunkmatters.html (April 1, 2012).

7. Morison, George Shattuck. *The New Epoch and the University* (Oration, delivered before the Phi Beta Kappa Society in Sanders Theatre, Cambridge, Mass., June 25, 1896). (3)

8. Ibid. (6)

9. Morison, George Shattuck. *The New Epoch.* (1896) Boston: Houghton, Mifflin and Co., 1903. (46)

10. Berthelot, Marcellin, "En l'an 2000," *Science et morale.* Paris: Lévy, 1897. (508–515). An English translation has been published: "In the Year 2000," *Chemistry in Britain,* Vol. 15 (1979). (250–251); translations by Alan Rocke, to whom I am indebted for the anecdote.

11. Morison, George Shattuck. *The New Epoch.* (1896) Boston: Houghton, Mifflin and Co., 1903. (46)

12. Morison, George Shattuck. *The New Epoch and the University* (Oration, delivered before the Phi Beta Kappa Society in Sanders Theatre, Cambridge, Mass., June 25, 1896). (4)

13. Yarow, Jay. "Meet the Startup That's Going to Make Eggs (Yes, Eggs) Obsolete," *Business Insider.* Dec. 7, 2013.

14. Dickens, Charles. *Hard Times.* Eds. Fred Kaplan and Sylvere Monod. New York: Norton and Company, 2001. (20–21)

15. Franklin, H. Bruce, ed. *Future Perfect: American Science Fiction of the Nineteenth Century: An Anthology.* (Expanded and revised ed. New Brunswick, N.J.: Rutgers University Press, 1995). (131)

16. Quoted in Matus, Jill L. *Shock, Memory and the Unconscious in Victorian Fiction.* 1st ed. Cambridge, England: Cambridge University Press, 2009. (83)

17. Guedalla, Philip. *Supers and Supermen: Studies in Politics, History and Letters.* New York: Knopf, 1921. (33)

18. Morison, Elting, *Men, Machines, and Modern Times.* Cambridge, Mass.: MIT Press, 1966. (6–7)

19. Morison, George Shattuck. *The New Epoch and the University* (Oration, delivered before the Phi Beta Kappa Society in Sanders Theatre, Cambridge, Mass., June 25, 1896). (7)

20. Morison, Elting. *Men, Machines, and Modern Times.* Cambridge, Mass.: MIT Press, 1966.

PART ONE: CHAOS • ORDER

1. Letter to Herwart von Hohenburg (Feb. 10, 1605); quoted in Holton, *Johannes Kepler's Universe: Its Physics and Metaphysics,* 342, as cited by Hylarie.

ONE: THE GOD OF MATHEMATICS

1. Aristotle ca. 355 B.C.E. *On the Heavens* I, 270b1; quoted in John Deely. *Four Ages of Understanding.* Toronto: University of Toronto Press, 2001. (79)

2. Wootton, David. *The Invention of Science.* New York: Harper, 2015. (3)

3. Morison, George Shattuck. *The New Epoch and the University* (Oration, delivered before the Phi Beta Kappa Society in Sanders Theatre, Cambridge, Mass., June 25, 1896). (1)

4. Wootton, David. *The Invention of Science.* New York: Harper, 2015. (4)

5. Ibid. (4)

6. Ibid. (1)

7. Morison, George Shattuck. *The New Epoch*. (1896) Boston: Houghton, Mifflin and Co., 1903. (130)

8. Power, Henry. *Experimental Philosophy* (1664), quoted in Wootton. (1)

9. Wootton, David. *The Invention of Science*. New York: Harper, 2015. (74)

10. Dolnick, Edward. *Clockwork Universe*. New York: Harper, 2011.

11. Wootton, David. *The Invention of Science*. New York: Harper, 2015. (6)

12. Ibid. (13)

13. Peters, S. M. (2008-02-05). *Whitechapel Gods* (Kindle locations 5435–5436). Penguin Publishing Group. Kindle Edition.

14. Morison, Arthur. *Palace Journal*. April 24, 1889.

15. Morison, Elting. *Men, Machines, and Modern Times*. Cambridge, Mass.: MIT Press, 1966. (9)

16. Wootton, David. *The Invention of Science*. New York: Harper, 2015. (217)

17. Quoted in Dolnick, Edward. *Clockwork Universe*. New York: Harper, 2011. (171)

18. Ibid. (146)

19. Translation quoted from Barker, Peter, and Bernard R. Goldstein. "Theological Foundations of Kepler's Astronomy." *Osiris*, Vol. 16, *Science in Theistic Contexts: Cognitive Dimensions*. Chicago: University of Chicago Press, 2001. (92)

20. Translation quoted from Barker, Peter, and Bernard R. Goldstein. "Theological Foundations of Kepler's Astronomy." (99)

21. Elliott, Rebecca. "Mysterium Cosmographicum: How to Make Your Own Platonic Solids," *PioneerTown*, 2015. Access June 16, 2016.

22. Caspar, Max, and Clarisse Doris Hellman. *Kepler*. Dover Books, 1993. (71)

23. Barker, Peter, and Bernard R. Goldstein. "Theological Foundations of Kepler's Astronomy." (102)

24. Ibid. (88)

25. Kepler, Johannes. *Kepler's Somnium: The Dream, or Posthumous Work on Lunar Astronomy,* trans. Edward Rosen. Courier Corporation, 1967. For an excellent summary and analysis, see Christianson, Gale. "Kepler's Somnium: Science Fiction and the Renaissance Scientist," *Science Fiction*, Vol. 8, no. 3 (1976).

26. Herschel, William, quoted in Holmes, Richard. *The Age of Wonder*. New York: Vintage Books, 2008. (62)

27. Dolnick, Edward. *Clockwork Universe*. New York: Harper, 2011. (231)

28. Djerassi, Carl, and David Pinner. *Newton's Darkness: Two Dramatic Views*. London: Imperial College Press, 2003. (3)

29. Ibid. (3)

30. Dodds, Betty Jo Teeter, quoted in Ramati, Ayval. "The Hidden Truth of Creation: Newton's Method of Fluxions," *The British Journal for the History of Science*, Vol. 34, No. 4 (Dec. 2001). (417–438)

31. The 87th recipe of the Leiden Papyrus, translated by Principe, Lawrence. *The Secrets of Alchemy*. Chicago: University of Chicago Press, 2013. (10)

32. Djerassi, Carl, and David Pinner. *Newton's Darkness: Two Dramatic Views*. London: Imperial College Press, 2003. (3)

33. Ramati, Ayval. "The Hidden Truth of Creation: Newton's Method of Fluxions," *The British Journal for the History of Science*, Vol. 34, No. 4 (Dec. 2001). (418)

34. Quoted from Westfall, Richard. "Newton's Marvelous Years of Discovery and Their Aftermath: Myth Versus Manuscript" *Isis*, Vol. 71, No. 1 (Mar. 1980). (109–121)

35. Ibid. (109)

36. Ibid. (109)

37. From Zakai, Avihu. *Jonathan Edwards's Philosophy of Nature: The Re-enchantment of the World in the Age of Scientific Reasoning*. London: Bloomsbury T&T Clark, 2010. (190)

38. Dolnick, Edward. *Clockwork Universe*. New York: Harper, 2011. (238)

39. Ibid. (257)

40. Ibid. (257)

41. Quoted from Tasaday, Nicholas. "When Lions Battle." *Math Horizons,* Vol. 14, No. 4 (April 2007). (8–11, 30–31)

42. Muller, Martin. "The Medieval Astronomical Clock in Prague," Prague.cz. Web. 2013. Accessed Feb. 9, 2015.

43. Wiener, Philip P. ed. *Leibniz Selections* (New York: Charles Scribner's Sons, 1951). (216–217)

44. Drake, Judith. *An Essay in Defence of the Female Sex. In a letter to a lady. Written by a lady*. 4th ed., corrected. London, 1721. Eighteenth Century Collections Online. Gale. Case Western Reserve University. (13)

45. Peters, S. M. *Whitechapel Gods* (Kindle locations 5440–5443). Penguin Publishing Group. Kindle Edition.

46. Ibid.

47. Newton, Isaac. "The Commentary on the Emerald Tablet," *The Alchemy Reader: From Hermes Trismegistus to Isaac Newton*. Ed. Stanton J. Linden. Cambridge, England: Cambridge University Press, 2003. (247)

48. Morison, Elting. *Men, Machines, and Modern Times*. Cambridge, Mass.: MIT Press, 1966. (206)

**TWO: CLOCKWORK BOY AND THE MOTHER MACHINE**

1. Culpeper, Nicholas. *The English Physitian*. Enlarged edition. Cornhill, 1665. "To the Reader."

2. Ibid.

3. Ibid.

4. A comment originally made by Humphry Davy, whom we'll hear from again in part 2.

5. Riskin, Jessica. "Eighteenth-Century Wetware," *Representations*, Vol. 83 (Summer 2003). (97–125)

6. Wootton, David. *The Invention of Science*. New York: Harper, 2015. (432)

7. Descartes, René, *A Discourse on Method*. Orig. 1637. Project Gutenberg [Ebook #59]. Produced by Ilana and Greg Newby. HTML version by Al Haines. July 1, 2008. (NP)

8. Ibid., sec 4.

9. Ibid., sec 2.

10. Ibid., sec 5.

11. Wootton, David. *The Invention of Science*. New York: Harper, 2015. (377)

12. Ryle, Gilbert. *Concept of Mind* (1949), quoted in Wootton, David. (377)

13. Russon, Mary-Ann. "Tokyo: Humanoid Robot Woman ChihiraAico Directs Customers around Department Store," *International Business Times,* April 20, 2015. http://www.ibtimes.co.uk/tokyo-lifelike-humanoid-robot-woman-chihiraaico-directs-customers-around-department-store-video-1497340. Accessed July 31, 2015.

14. Algar, Jim. "Tokyo Museum Exhibit Boasts Remarkable Human-like Robots. Creepy or Cool?" *TechTimes.* June 24, 2015. http://www.techtimes.com/articles/9113/20140624/tokoyo-museum-boasts-super-human-like-robots.htm.

15. "Do Robots Have Souls?" Video interview. Hiroshi Ishiguro. *The Global Mail. Vimeo.* Accessed Aug. 7, 2015. https://vimeo.com/59110465.

16. Riskin, Jessica. "Eighteenth-Century Wetware," *Representations,* Vol. 83 (Summer 2003). (99)

17. Muri, Allison. *The Enlightenment Cyborg: A History of Communications and Control in the Human Machine, 1660–1830.* Toronto: University of Toronto Press, 2007. (37)

18. Ibid. (118)

19. d'Holbach, Baron, and Paul Henri Thierry. *System of Nature.* Sioux Falls, S.D.: NuVision Publications, LLC, Oct. 1, 2007. (62)

20. Berkeley, George. *Alciphron: Or. The Minute Philosopher. In Seven Dialogues.* J. Tonson in the Strand, 1732. (42)

21. Drake, Judith. *An Essay in Defence of the Female Sex. In a letter to a lady. Written by a lady.* The fourth edition, corrected. London, 1721. *Eighteenth Century Collections Online.* Gale. Case Western Reserve University. May 22, 2009. (13)

22. Selznick, Brian. *The Invention of Hugo Cabret*. New York: Scholastic Inc. September 15, 2015. Google ebook.

23. Dolnick, Edward. *Clockwork Universe*. New York: Harper, 2011. (8)

24. Sugg, Richard. *Mummies, Cannibals, and Vampires: The History of Corpse Medicine from the Renaissance to the Victorians*. London: Routledge, 2011. (1)

25. Burney, Fanny. "Account from Paris of a Terrible Operation—1812—" Journal Letter to Esther Burney, March 22–June 1812. *Fanny Burney. Selected Letters and Journals*. Ed. Joyce Hemlow. Oxford, England: Clarendon, 1986.

26. Peters, S. M. *Whitechapel Gods* (Kindle locations 147–149). New York: Penguin Publishing Group. Kindle Edition.

27. Tresch, John. "The Machine Awakens: The Science and Politics of the Fantastic Automaton," *French Historical Studies*, Vol. 34, No. 1 (Winter 2011).

28. Ibid. (7)

29. Riskin, Jessica. "Eighteenth-Century Wetware," *Representations*, Vol. 83 (Summer 2003). (97–125; 103)

30. Cottom, Daniel. "The Work of Art in the Age of Mechanical Digestion." *Representations*, Vol. 66, (Spring 1999). (56)

31. Voskuhl, Adelheid. *Androids in the Enlightenment: Mechanics, Artisans, and Cultures of the Self*. Chicago: University of Chicago Press, 2013. Google ebook. (NP)

32. Riskin, Jessica. "Eighteenth-Century Wetware," *Representations*, Vol. 83 (Summer 2003). (101)

33. Ibid. (102)

34. Hilaire-Pérez, Liliane, and Christelle Rabier. "Self-Machinery? Steel Trusses and the Management of Ruptures in Eighteenth-Century Europe," *Fitting for Health*. Society for the History of Technology, 2013. (460)

35. La Mettrie [1747]. Quotes and translations from: Niran Bahjat-Abbas. *Thinking Machines: Discourses of Artificial Intelligence*. LIT Verlag, 2006. (30)

36. Franchi, Stefano, and Güven Güzeldere. *Mechanical Bodies, Computational Minds: Artificial Intelligence from Automata to Cyborgs*. Cambridge, Mass.: MIT Press, 2005. (40)

37. Morison, Elting. *Men, Machines, and Modern Times*. Cambridge, Mass.: MIT Press, 1966. (9)

38. Wootton, David. *The Invention of Science*. New York: Harper, 2015. (364)

39. Hilaire-Pérez, Liliane, and Christelle Rabier. "Self-Machinery? Steel Trusses and the Management of Ruptures in Eighteenth-Century Europe," *Fitting for Health*. Society for the History of Technology, 2013. (471)

40. Ibid. (480)

41. Blackwell, Bonnie. "Tristram Shandy and the Theatre of the Mechanical Mother." *ELH*, Vol. 68, No. 1 (2001). (81–133)

42. Schillace, Brandy. "'Reproducing' Custom: Mechanical Habits and Female Machines in Augustan Women's Education," *Feminist Formations*, [Vol. 25, No. 1] (Spring 2013).

43. Blackwell, Bonnie. "Tristram Shandy." (87)

44. Astell, Mary. *A Serious Proposal to the Ladies*. [1694] Peterborough, Ont.: Broadview Press, 2002. (120)

45. Nihell, Elizabeth. "A Treatise on Midwifery." *Eighteenth-Century British Midwifery Volume 6*. London: Pickering and Chatto, 2008. (82)

46. Quoted in Johnstone, R. W. *William Smellie: The Master of British Midwifery*. Edinburgh: E & S Livingstone, 1952. (25)

47. Ibid. (26)

48. Quoted in King, Helen. *Midwifery, Obstetrics and the Rise of Gynaecology*. Farnham, England: Ashgate, 2007. (133)

49. Peters, S. M. *Whitechapel Gods* (Kindle locations 1637–1638). New York: Penguin Publishing Group. Kindle Edition.

50. Ibid.

51. Ibid.

52. Ibid.

53. Kang, Minsoo. *Sublime Dreams of Living Machines*. Cambridge, Mass.: Harvard University Press, 2011. (131)

54. Schillace, Brandy. "Reproducing." (10)

55. Nihell, Elizabeth. "A Treatise on Midwifery." *Eighteenth-Century British Midwifery Volume 6*. London: Pickering and Chatto, 2008. (82)

56. Morison, Elting. *Men, Machines, and Modern Times*. Cambridge, Mass.: MIT Press, 1966. (119)

57. Ibid. (83)

58. Ibid. (83)

THREE: CATCHING LIGHTNING IN A BOTTLE

1. Quoted in translation from Delbourgo, James. *A Most Amazing Scene of Wonders*. Cambridge, Mass.: Harvard University Press, 2006. (14)

2. Quoted in translation from ibid. (14).

3. Brundtland, Terje. "Francis Hauksbee and His Air Pump," *Notes and Records of the Royal Society*, Vol. 66 (2012). (255)

4. Hauksbee, Francis. *Physico-Mechanical Experiments on Various Subjects*. London, 1709. (Preface)

5.  Quoted in Bertucci, Paola. "Sparks in the Dark: The Attraction of Electricity in the Eighteenth Century," *Endeavour*, Vol. 31, No. 3. (NP)

6.  Quoted in ibid.

7.  Delbourgo, James. *A Most Amazing Scene of Wonders*. Cambridge, Mass.: Harvard University Press, 2006. (15)

8.  Bertucci, Paola. "Sparks in the Dark: The Attraction of Electricity in the Eighteenth Century," *Endeavour*, Vol. 31, No. 3. (2007).

9.  With thanks to the Bakken Museum, Minneapolis, Minn.

10. Delbourgo, James. *A Most Amazing Scene of Wonders*. Cambridge, Mass.: Harvard University Press, 2006. (36)

11. Franklin, Benjamin. Letter to Collinson quoted in Delbourgo, James. *A Most Amazing Scene of Wonders*. Cambridge, Mass.: Harvard University Press, 2006. (38)

12. Quoted in Bertucci, Paola. "Sparks in the Dark: The Attraction of Electricity in the Eighteenth Century," *Endeavour*, Vol. 31, No. 3. (NP)

13. Ibid.

14. Shelley, Mary Wollstonecraft. *Frankenstein; or the Modern Prometheus*. Project Gutenberg. Release date: June 17, 2008 [Ebook #84]. (NP)

15. Delbourgo, James. *A Most Amazing Scene of Wonders*. Cambridge, Mass.: Harvard University Press, 2006. (3)

16. Crabtree, S. (producer), and Jim Al-Khalili (presenter). *Shock and Awe: The Story of Electricity*. [Documentary] United Kingdom: BBC.

17. Franklin, Benjamin, quoted in Finger, Stanley. "Benjamin Franklin and the Electrical Cure." *Brain, Mind and Medicine: Essays in Eighteenth-Century Neuroscience*. Eds. Harry Whitaker, C.U.M. Smith, and Stanley Finger. (246)

18. Finger, Stanley. "Benjamin Franklin and the Electrical Cure." *Brain, Mind and Medicine: Essays in Eighteenth-Century Neuroscience*. Eds. Harry Whitaker, C.U.M. Smith, and Stanley Finger. (246; 271–283)

19. Shelley, Mary Wollstonecraft. *Frankenstein; or the Modern Prometheus*. Project Gutenberg. Release date: June 17, 2008 [Ebook #84] (NP)

20. Ibid.

21. Delbourgo, James. *A Most Amazing Scene of Wonders*. Cambridge, Mass.: Harvard University Press, 2006. (9)

22. Bertucci, Paola. "Therapeutic Electricity." *Brain, Mind and Medicine: Essays in Eighteenth-Century Neuroscience*. Eds. Harry Whitaker, C.U.M. Smith, and Stanley Finger. (271–283)

23. *Epitome of Electricity and Galvanism* 1809; quoted in Delbourgo, James. *A Most Amazing Scene of Wonders*. Cambridge, Mass.: Harvard University Press, 2006. (8)

24. Shelley, Mary Wollstonecraft. *Frankenstein; or the Modern Prometheus*. Project Gutenberg. Release date: June 17, 2008 [Ebook #84]. (NP)

25. Turkel, William. *Sparks from the Deep: How Shocking Experiments with Strongly Electrical Fish Powered Scientific Discovery*. Baltimore: Johns Hopkins University Press, 2013. (80)

26. Shelley, Mary Wollstonecraft. *Frankenstein; or the Modern Prometheus*. Project Gutenberg. Release date: June 17, 2008 [Ebook #84]. (NP)

27. Fowler, Richard. *Experiments and Observations relative to the influence lately discovered by Galvani and commonly called Animal Electricity*. Edinburgh, 1793. (1)

28. Ibid.

29. Shelley, Mary Wollstonecraft. *Frankenstein; or the Modern Prometheus*. Project Gutenberg. Release date: June 17, 2008 [Ebook #84]. (NP)

30. *Intellectual Electricity, Novum Organum of Vistion, and Grand Mystic Secret* (Anon.) London, 1798.

31. Shelley, Mary Wollstonecraft. *Frankenstein; or the Modern Prometheus*. Project Gutenberg. Release date: June 17, 2008 [Ebook #84]. (NP)

32. Ibid.

33. Montillo, Roseanne. *The Lady and Her Monsters*. New York: William Morrow, 2013. (54)

34. Holmes, Richard. *The Age of Wonder*. New York: Vintage Books, 2008. (125).

35. Ibid. (132)

36. Turkel, William. *Sparks from the Deep: How Shocking Experiments with Strongly Electrical Fish Powered Scientific Discovery*. Baltimore: Johns Hopkins University Press, 2013. (80)

37. Pancaldi, Giuliano. *Volta: Science and Culture in the Age of Enlightenment*. Princeton, N.J.: Princeton University Press, 2005. (180).

38. Turkel, William. *Sparks from the Deep: How Shocking Experiments with Strongly Electrical Fish Powered Scientific Discovery*. Baltimore: Johns Hopkins University Press, 2013. (81)

39. Shelley, Mary Wollstonecraft. *Frankenstein; or the Modern Prometheus*. Project Gutenberg. Release date: June 17, 2008 [Ebook #84]. (NP)

40. *An Account of the Late Improvements in Galvanism*, quoted in Holmes, Richard. *The Age of Wonder*. New York: Vintage Books, 2008. (317)

41. Ibid. (328)

42. Ibid. (329)

43. Ibid. (192)

44. Darwin, Erasmus, quoted in Holmes, Richard. *The Age of Wonder*. New York: Vintage Books, 2008. (197)

# ENDNOTES

45. Ibid. (198)

46. The best description of Herschel's incredible discoveries, as well as the lasting impact they had on our modern future, can be found in Richard Holmes, *Age of Wonder.*

47. Byron, George Gordon. Letters. *"Alas! The Love of Women": 1813–1814.* Ed. Leslie Alexis Marchand. Cambridge, Mass: Harvard University Press, 1974. (64)

48. Holmes, Richard. *The Age of Wonder.* New York: Vintage Books, 2008. (210)

49. Ibid. (235)

50. Taylor, W. C. *The Modern British Plutarch; of Lives of Men Distinguished in The Recent History of England.* New York, 1868. (113)

51. Ibid. (113)

52. Ibid. (114–115)

53. Davy, Humprhy, quoted in Holmes, Richard. *The Age of Wonder.* New York: Vintage Books, 2008. (254)

54. Davy, Humprhy, quoted in ibid. (254)

55. Shelley, Mary Wollstonecraft. *Frankenstein; or the Modern Prometheus.* Project Gutenberg. Release date: June 17, 2008 [Ebook #84]. (NP)

56. Holmes, Richard. *The Age of Wonder.* New York: Vintage Books, 2008. (260)

57. Davy, Humphry, to J. King Esquire. *Fragmentary Remains, Literary and Scientific, of Sir Humphry Davy.* Ed. J. Churchill. London. 1858. (64)

58. Davy, Humphry, quoted in Holmes, Richard. *The Age of Wonder.* New York: Vintage Books, 2008. (289)

59. Tuft, Michael, and Victoria Bell (producers), and Michael Mosely, Mark Miodownik, and Cassie Newland (presenters). *Genius of Invention.* Episode 1. [Documentary] United Kingdom: BBC.

60. Boswell, James. *The Life of Samuel Johnson, L.L.D.* Edinburgh. 1890. (281)

61. Davy, Humphry. *The Collected Works of Sir Humphry Davy (with Miscellaneous Papers and Researches, especially on The Safety Lamp, and Flame).* Ed. John Davy. London, 1840. (5)

62. Davy, John. *Memoirs of the Life of Sir Humphry Davy.* London. 1836. (1)

63. Blum, Deborah. "Firedamp," *Speakeasy Science,* ScienceBlogs. April 9, 2010. http://scienceblogs.com/speakeasyscience/2010/04/09/firedamp/.

64. Davy, Humphry. *The Collected Works of Sir Humphry Davy (with Miscellaneous Papers and Researches, especially on The Safety Lamp, and Flame).* Ed. John Davy. London, 1840. (5)

65. Ibid. (5)

66. "Felling" Mining Disasters—Names. Durham Mining Museum. http://www.dmm.org.uk/names/n1812-01.htm.

67. Holmes, Richard. *The Age of Wonder*. New York: Vintage Books, 2008. (363)

68. Davy, Humphry. *The Collected Works of Sir Humphry Davy (with Miscellaneous Papers and Researches, especially on The Safety Lamp, and Flame)*. Ed. John Davy. London, 1840. (10)

69. Ibid. (11)

70. Holmes, Richard. *The Age of Wonder*. New York: Vintage Books, 2008. (365)

71. Ibid. (367)

72. Shelley, Mary Wollstonecraft. *Frankenstein; or the Modern Prometheus*. Project Gutenberg. Release date: June 17, 2008 [Ebook #84]. (NP)

73. Davy, Humphry. *The Collected Works of Sir Humphry Davy (with Miscellaneous Papers and Researches, especially on The Safety Lamp, and Flame)*. Ed. John Davy. London, 1840 (18)

74. Buddle, John. Report of the Select Committee, 1835. As quoted in Holmes, Richard. *The Age of Wonder*. New York: Vintage Books, 2008. (368)

75. Ibid. (369)

76. Abernathy, John. *An Enquiry into the Probability and Rationability of Mr. Hunter's Theory of Life*. London. 1814. (48)

77. Ibid. (49)

78. Shelley, Mary Wollstonecraft. *Frankenstein; or the Modern Prometheus*. Project Gutenberg. Release date: June 17, 2008 [Ebook #84]. (NP).

79. Davy, John. *Memoirs of the Life of Sir Humphry Davy*. London, 1836. (378–379)

80. Ibid. (379)

81. Ibid. (380)

82. Ibid. (385)

83. Shelley, Mary Wollstonecraft. *Frankenstein; or the Modern Prometheus*. Project Gutenberg. Release date: June 17, 2008 [Ebook #84]. (NP)

84. Holmes, Richard. *The Age of Wonder*. New York: Vintage Books, 2008. (430)

**FOUR: INTO DARK COUNTRY**

1. Shelley, Mary Wollstonecraft. *Frankenstein; or the Modern Prometheus*. Project Gutenberg. Release date: June 17, 2008 [Ebook #84]. (NP)

2. Morison, Elting. *Men, Machines, and Modern Times*. Cambridge, Mass.: MIT Press, 1966. (18)

3. Morison, George Shattuck. *The New Epoch and the University* (Oration, delivered before the Phi Beta Kappa Society in Sanders Theatre, Cambridge, Mass., June 25, 1896).

4. Wallace, Willard Mosher. *Sir Walter Raleigh*. Princeton, N.J.: Princeton University Press, 2015. (120)

5.  Magasich-Airola, Jorge, and Jean-Marc de Beer. *America Magica,* 2nd ed. *When Renaissance Europe Thought It Had Conquered Paradise.* London: Anthem Press, 2007. (4)

6.  Long, Edward, *History of Jamaica* (London: 1774), Vol. II. Quoted in Giraldo, Alexander. "Obeah: The Ultimate Resistance," *Slave Resistance, a Caribbean Study.* http://scholar.library.miami.edu/slaves/Religion/religion.html.

7.  Williams, Joseph J., *Voodoos and Obeahs: Phases of West India Witchcraft* (New York: Dial Press, 1932). Quoted in Giraldo, Alexander. "Obeah: The Ultimate Resistance," *Slave Resistance, a Caribbean Study.* http://scholar.library.miami.edu/slaves/Religion/religion.html.

8.  Evans, Arthur B. *Jules Verne Rediscovered: Didacticism and the Scientific Novel.* Westport, Conn.: Greenwood Publishing Group, 1988. (21)

9.  Ibid. (17)

10. Verne, Jules. *Great Navigators of the Eighteenth Century.* [1880] London, 2001. (3)

11. Ibid. (29)

12. Ibid. (30)

13. Quoted in ibid. (35)

14. Ibid. (36)

15. Ibid. (55)

16. Ibid. (58)

17. Holmes, Richard. *The Age of Wonder.* New York: Vintage Books, 2008. (1)

18. Banks, Joseph, quoted in ibid. (3)

19. Quoted in Verne, Jules. *Great Navigators of the Eighteenth Century.* [1880] London, 2001. (63)

20. Holmes, Richard. *The Age of Wonder.* New York: Vintage Books, 2008. (6)

21. Ibid. (7)

22. Quoted in ibid.

23. Terry, Martin, and Susan Hall. *Cook's Endeavour Journal: The Inside Story.* Canberra: National Library of Australia, 2008. (20)

24. *London Gazette,* Aug. 18, 1768. Quoted in Terry, Martin, and Susan Hall. Ibid. (23)

25. Banks, Joseph, quoted in Holmes, Richard. *The Age of Wonder.* New York: Vintage Books, 2008. (11)

26. Verne, Jules. *Great Navigators of the Eighteenth Century.* [1880] London, 2001. (114)

27. Ibid. (128)

28. Cook, James, quoted in ibid. (128)

29. Coleridge, Samuel. *Rime of the Ancient Mariner.* Part the First. Project Gutenberg. Release date: March 11, 2006 [Ebook #151]. Last updated: January 26, 2013. (NP)

30. Cook, James. *A Voyage Towards the South Pole and Round the World V 1.* Project Gutenberg. Contributor: Tobias Furneaux. Release date: May 7, 2005 [Ebook #15777]. (NP)

31. Ibid.

32. Krebs, Robert. *The Basics of Earth Science.* Westport, Conn.: Greenwood Publishing, 2003. (140)

33. Frank, Frederick, and Diane Long Hoeveler, eds. "Introduction." Edgar Allan Poe. *The Narrative of Arthur Gordon Pym of Nantucket.* Peterborough, Ont.: Broadview Editions, 2010. (24)

34. Meville, Herman. *Moby-Dick; Or, The Whale.* Posting date: December 25, 2008 [Ebook #2701]. Release date: June 2001. (NP)

35. "Doomed Expedition to the Pole, 1912," *EyeWitness to History,* www.eyewitnessto history.com (1999).

36. Ibid.

37. Delgado, James. *Across the Top of the World: The Quest for the Northwest Passage,* Madeira Park, B.C.: Douglas & McIntyre, 2009. (xiii)

38. Elce, Erika Behrisch. "'In Sum, Evil Has Prevailed': The Moral Morass of Science and Exploration in Jacques Tardi's *The Arctic Marauder.*" *Steaming into a Victorian Future.* Eds. Julie Anne Taddeo and Cynthia J. Miller. New York: Rowman and Littlefield, 2014. (106)

39. Delgado, James. *Across the Top of the World: The Quest for the Northwest Passage,* Madeira Park, B.C.: Douglas & McIntyre, 2009. (xiii)

40. Holmes, Richard. *The Age of Wonder.* New York: Vintage Books, 2008. (211)

41. Banks, Sir Joseph. Letter, 1803. Quoted in ibid. (211)

42. McLynn, Frank. *Hearts of Darkness: The European Exploration of Africa.* Quoted in McCarthy, James. *Journey into Africa: The Life and Death of Keith Johnson, Scottish Cartographer and Explorer.* Dunbeath Mill, England: Whittles Publishing Group, 2004. (144)

43. Holmes, Richard. *The Age of Wonder.* New York: Vintage Books, 2008. (212)

44. Ibid. (212)

45. Park, Mungo, quoted in Verne, Jules. *Great Navigators of the Eighteenth Century.* [1880] London, 2001. (392)

46. Holmes, Richard. *The Age of Wonder.* New York: Vintage Books, 2008. (217)

47. Conrad, Joseph. "Geography," 1924, quoted in ibid. (220)

48. Conrad, Joseph. *Heart of Darkness.* Project Gutenberg. Release date: January 9, 2006 [Ebook #526]. (NP)

49. Verne, Jules. *Great Navigators of the Eighteenth Century.* [1880] London, 2001. (404)

50. MacDonald, George. *Unspoken Sermons.* New York: Cosimo Books, 2007. (248)

51. Morison, George Shattuck. *The New Epoch and the University* (Oration, delivered before the Phi Beta Kappa Society in Sanders Theatre, Cambridge, Mass., June 25, 1896). (27)

FIVE: THE SCIENTIST AND THE ENGINEER

1. Shelley, Mary Wollstonecraft. "Preface, 1831 edition." *Frankenstein, Or the Modern Prometheus.* Ware, England: Wordsworth Editions, 1993. (5)

2. Csicsery-Ronay, Istvan, Jr. *The Seven Beauties of Science Fiction.* Middletown, Conn.: Wesleyan University Press, 2011. (NP)

3. Morison, George Shattuck. *The New Epoch.* (1896) Boston: Houghton, Mifflin and Co., 1903. (8)

4. Ibid. (8)

5. Ibid. (107)

6. Ibid. (16)

7. Ibid. (16)

8. Holmes, Richard. *The Age of Wonder.* New York: Vintage Books, 2008. (406)

9. Quoted in Collier, Bruce, and James MacLachlan. *Charles Babbage: And the Engines of Perfection.* New York: Oxford University Press, 1998. (9)

10. Quoted in ibid. (9)

11. A commentator to Davy's voltaic pile, 1800, quoted in Cantor, Geoffrey. *Michael Faraday: Sandemanian and Scientist.* New York: Macmillan, 1991. (140)

12. Holmes, Richard. *The Age of Wonder.* New York: Vintage Books, 2008. (403)

13. Pollock, Lady Jane. *St. Paul's Magazine.* Quoted in Morus, Iwan Rhys. *Shocking Bodies: Life, Death, and Electricity in Victorian England.* Stroud, England: The History Press, 2011. (71)

14. Schwarz, K. K. "Faraday and Babbage," *Notes and Records of the Royal Society of London* Vol. 56, No. 3 (Sept. 2002). (367–381)

15. Holmes, Richard. *The Age of Wonder.* New York: Vintage Books, 2008. (436)

16. Ibid. (440)

17. Ibid. (441)

18. Morus, Iwan Rhys. *Shocking Bodies: Life, Death, and Electricity in Victorian England.* Stroud, England: The History Press, 2011. (58)

19. Essinger, James. *Ada's Algorithm: How Lord Byron's Daughter Ada Lovelace Launched the Digital Age.* New York: Melville House, 2014. Ebook. (Chap 10, NP)

20. Morus, Iwan Rhys. *Shocking Bodies: Life, Death, and Electricity in Victorian England.* Stroud, England: The History Press, 2011. (62)

21. Lovelace, Ada, quoted in ibid. (62)

22. Babbage, Charles. Letter. Quoted in *Ada's Algorithm: How Lord Byron's Daughter Ada Lovelace Launched the Digital Age.* New York: Melville House, 2014. Ebook. (Chap. 13, NP)

23. Lovelace, Ada. "Notes by the Translator," for Menabrea, L. F. "Sketch of the Analytical Engine by Charles Babbage, Esq.," *Scientific Memoirs* 3, 1843. (666–731)

24. George Zarkadakis, *In Our Own Image: Savior or Destroyer? The History and Future of Artificial Intelligence.* New York: Pegasus Books, 2016.

25. Padua, Sydney, quoted in Lehoczky, Etelka. "Lovelace and Babbage Is a Thrilling Adventure," NPR Book Reviews. April 23, 2015.

26. Morus, Iwan Rhys. *Shocking Bodies: Life, Death, and Electricity in Victorian England.* Stroud, England: The History Press, 2011. (63)

27. Lovelace, Ada, quoted in ibid. (68)

28. Gibson, William, and Bruce Sterling. *The Difference Engine.* New York: Random House, 2011. (26)

29. Morison, Elting. *Men, Machines, and Modern Times.* Cambridge, Mass.: MIT Press, 1966. (99)

30. Ibid. (115)

31. Quoted in ibid. (115)

32. Ibid. (119)

33. Lovelace, Ada. "Notes by the Translator," for Menabrea, L. F. "Sketch of the Analytical Engine by Charles Babbage, Esq." *Scientific Memoirs* 3, 1843. (666–731)

34. Morus, Iwan Rhys. *Shocking Bodies: Life, Death, and Electricity in Victorian England.* Stroud, England: The History Press, 2011. (65)

35. Description of Crosse estate, quoted in ibid. (64)

36. Ibid. (70)

37. Sopwith, Thomas, quoted in Heald, Henrietta. *William Armstrong: Magician of the North.* Carmarthen, Wales: McNidder and Grace, 2012. (286)

38. Quoted in ibid. (3)

39. The *Times*, 1884; quoted in ibid. (5)

40. Gibson, Thomas George, Henry Clapham, and Hill Morum. *The Proceedings and reports of the town council of the borough of Newcastle. Proceedings of the council of the city and county of Newcastle-upon-Tyne.* 1884. (xl)

41. Heald, Henrietta. *William Armstrong: Magician of the North.* Carmarthen, Wales: McNidder and Grace, 2012. (14)

42. Morison, George Shattuck. *The New Epoch.* [1896] Boston: Houghton, Mifflin and Co., 1903. (67)

43. Member of Parliament Joseph Cowen, quoted in Heald, Henrietta. *William Armstrong: Magician of the North*. McNidder and Grace, 2012. (8)

44. Heald, Henrietta. *William Armstrong: Magician of the North*. Carmarthen, Wales: McNidder and Grace, 2012. (39)

45. J. W. Richardson quoted in ibid. (9)

46. Ibid. (38)

47. Ibid. (43)

48. Sopwith, Thomas, ed. Sir Benjamin Ward Richardson. *Thomas Sopwith: With Excerpts from his Diary of Fifty-Seven Years*. Longmans Green, 1891. (232)

49. Mayhew, Henry. *London Labour and the London Poor*. New York: Penguin Classics, 1985. (371)

50. "Polluted State of the Thames," *The Mechanics Magazine*, Vol. 63, 1855. (28)

51. Bell, Thomas J. *History of the Water Supply of the World: Arranged in a Comprehensive Form [. . .]*, 1882. (21)

52. Heald, Henrietta. *William Armstrong: Magician of the North*. Carmarthen, Wales: McNidder and Grace, 2012. (57)

53. Warren, Kenneth. *Armstrongs of Elswick: Growth in Engineering and Armaments*. New York: Springer, 1989. (11)

54. Melville, Herman. *Moby-Dick*. New American Library, 1892. Google ebook (391).

55. Armstrong, William, quoted in Heald, Henrietta. *William Armstrong: Magician of the North*. Carmarthen, Wales: McNidder and Grace 2012. (62)

56. Armstrong, William, quoted in ibid. (62)

57. Armstrong, William, quoted in ibid. (77)

58. "Monster Gun." *William Armstrong* website. iTough Ltd. Sept. 2, 2016. http://www.williamarmstrong.info/monster-gun.

59. Padua, Sydney. Interview, Aug. 27, 2016.

60. Collier, Bruce. *The Little Engines That Could've*. New York: Garland, 1990. C4, ebook. (NP)

61. Lord Moulton, quoted in ibid. (NP)

62. "Address of Sir W. Armstrong, Before the Mechanical Science Section," *The Electrical Review*, Vol. 9 (Sept. 15, 1881). (349)

63. Ibid. (350–351)

**SIX: HE WHO POWERS THE FUTURE**

1. Barth, Linda. *A History of Inventing in New Jersey: From Thomas Edison to the Ice Cream Cone*. Mt. Pleasant, S.C.: Arcadia Press. 2013.

2. Jonnes, Jill. *Empires of Light*. New York: Random House. 2003. (52)

3.  Edison, quoted in ibid. (52)

4.  *New York Daily Graphic*, quoted in ibid. (54)

5.  Morison, George Shattuck. *The New Epoch and the University*, (Oration, delivered before the Phi Beta Kappa Society in Sanders Theatre, Cambridge, Mass., June 25, 1896. (17)

6.  Jonnes, Jill. *Empires of Light*. New York: Random House. 2003. (54)

7.  Ibid. (56)

8.  Edison, quoted in ibid. (56)

9.  *New York Daily Graphic*, quoted in ibid. (54)

10. Ibid. (57)

11. *New York Sun,* quoted in ibid. (57)

12. Tesla, Nikola, quoted in ibid. (105)

13. Ibid. (10)

14. Seifer, Marc J. *Wizard: The Life and Times of Nikola Tesla: Biography of a Genius.* Secaucus, N.J.: Citadel Press, 1998. (11)

15. Tesla, Nikola. *My Inventions and Other Writings.* Courier Dover Publications. (25)

16. Tesla, Nikola, quoted in Seifer, Marc J. *Wizard: The Life and Times of Nikola Tesla: Biography of a Genius.* Secaucus, N.J.: Citadel Press, 1998. (71)

17. O'Neill, James. *Prodigal Genius: The Life of Nikola Tesla.* New material and revisions, New York: Cosimo Classics. 2007. (3)

18. Tesla, Nikola, quoted in Seifer, Marc J. *Wizard: The Life and Times of Nikola Tesla: Biography of a Genius.* Secaucus, N.J.: Citadel Press, 1998. (71)

19. Ibid. (85)

20. Winter, Alison. *Mesmerism.* Chicago: University of Chicago Press, 1998. (1)

21. Crookes, William. *Researches in the Phenomena of Spiritualism.* London: J. Burns, 1874. (4)

22. Hughes, Jeff. "A Crookes Charter." Review of *William Crookes (1832–1919) and the commercialization of science* by W. H. Brock. *Notes and Records of the Royal Society of London*, Vol. 64, No. 1 (March 20, 2010). (91–93)

23. Crookes, William. *Researches in the Phenomena of Spiritualism.* London: J. Burns. 1874.

24. Quoted in Morus, Iwan Rhys. *Shocking Bodies: Life, Death, and Electricity in Victorian England.* Stroud, England: The History Press, 2011. (113–114)

25. Quoted in ibid. (114).

26. Bulwer-Lytton, Edward. *Vril, the Power of the Coming Race* (Kindle locations 119–121). Chasma Press. Kindle Edition.

27. Ibid.

28. Ibid.

29. Quoted in Seifer, Marc J. *Wizard: The Life and Times of Nikola Tesla: Biography of a Genius*. Secaucus, N.J.: Citadel Press, 1998. (63)

30. Ibid. (64)

31. "New IOM Members," *Science*, Vol. 214, No. 4520 (1981). http://www.jstor.org/stable/1687288. (524)

32. Quoted in Seifer, Marc J. *Wizard: The Life and Times of Nikola Tesla: Biography of a Genius*. Secaucus, N.J.: Citadel Press, 1998. (196)

33. Bulwer-Lytton, Edward. *Vril, the Power of the Coming Race* (Kindle location 447). Chasma Press. Kindle Edition.

34. Tesla, Nikola. *My Inventions and Other Writings*. Courier Dover Publications. (23)

35. Seifer, Marc J. *Wizard: The Life and Times of Nikola Tesla: Biography of a Genius*. Secaucus, N.J.: Citadel Press, 1998. (16)

36. Quoted in ibid. (16)

37. Tesla, Nikola. *My Inventions and Other Writings*. Courier Dover Publications. (27)

38. Prichard, James Cowles. "Forms of Insanity." Excerpted. *Embodied Selves: An Anthology of Psychological Texts 1830–1890*. Eds. Taylor, Jenny Bourne, and Sally Shuttleworth. Oxford, England: Clarendon Press, 2003. (252)

39. Esquirol, Etienne. *Mental Maladies; a Treatise on Insanity*. Trans. E. K. Hunt. Philadelphia: Lea and Blanchard, 1845. (200)

40. Tesla, Nikola. *My Inventions and Other Writings*. Courier Dover Publications. (26)

41. Ibid. (28)

42. Seifer, Marc J. *Wizard: The Life and Times of Nikola Tesla: Biography of a Genius*. Secaucus, N.J.: Citadel Press, 1998. (23)

43. Tesla, Nikola, quoted in Jonnes, Jill. *Empires of Light*. New York: Random House. 2003. (93)

44. Macomber, William. *The Fixed Law of Patents: As Established by the Supreme Court of the United States and the Nine Circuit Courts of Appeals*. New York: Little, Brown, 1913. (671)

45. Seifer, Marc J. *Wizard: The Life and Times of Nikola Tesla: Biography of a Genius*. Secaucus, N.J.: Citadel Press, 1998. (54)

46. Ibid. (56)

47. Quoted in ibid. (58)

48. Seifer, Marc J. *Wizard: The Life and Times of Nikola Tesla: Biography of a Genius*. Secaucus, N.J.: Citadel Press, 1998. (78)

49. Ibid. (79)

50. Tesla, Nikola. *Inventions, Researches, and Writings of Nikola Tesla*. Ed. Martin, T. C. 1894. (318)

51. Ibid. (319)

52. Childress, David Hatcher. *The Tesla Papers*, 162.

53. Jonnes, Jill. *Empires of Light*. New York: Random House. 2003. (316)

54. Tesla, Nikola. "The Age of Electricity," *Cassier's Magazine: An Engineering Monthly*. Vol. 11, 1897. Ebook, Google. (385)

55. Tesla, Nikola. "Nikola Tesla and His Wonderful Discoveries." *Electrical World*, April 29, 1893. (324)

56. Steltzner, Adam. "The Great Math Mystery," *Nova* (2015).

57. Nolan, Christopher. *The Prestige*. 2006.

58. Ibid.

59. Letter by Nikola Tesla, quoted in Seifer, Marc J. *Wizard: The Life and Times of Nikola Tesla: Biography of a Genius*. Secaucus, N.J.: Citadel Press, 1998. (328)

60. Morison, Elting. *Men, Machines, and Modern Times*. Cambridge, Mass.: MIT Press, 1966. (119)

SEVEN: A WRENCH IN THE AGE OF MACHINERY

1. Heald, Henrietta. *William Armstrong: Magician of the North*. Carmarthen, Wales: McNidder and Grace, 2012. (45)

2. Martineau, Harriet. *Harriet Martineau's Autobiography, Abridged*. Big Byte Books. Google ebooks. (358)

3. Carlyle, Thomas. "Meeting of the British Association at Manchester," *Fraser's Magazine*, Vol. 26 (Sept. 1842). (366)

4. "Inundation of Heaton Colliery," *History of Mining in Durham and Northumberland*. Newcastle University. http://www.ncl.ac.uk/library/services/education-outreach/outreach/mining/heaton.php.

5. Sources compiled for "Child Labor during the British Industrial Revolution." *EH.net*. Economic History Association. (NP). https://eh.net/encyclopedia/child-labor-during-the-british-industrial-revolution/.

6. Carlyle, Thomas. *Sartor Resartus: The life and opinions of Herr Teufelsdröckh in three books*. Chapman and Hall, 1831. (46)

7. Carlyle, Thomas. "A Mechanical Age" (1829), first published in the *Edinburgh Review*.

8. Heald, Henrietta. *William Armstrong: Magician of the North*. Carmarthen, Wales: McNidder and Grace, 2012. (214)

9. Engels, Friedrich. *The Condition of the Working Class in Great Britain*. Oxford, England: Oxford World's Classics. (9)

10. Ibid. (9)

11. Ibid. (10)

12. Ibid. (17)

13. Morison, George Shattuck. *The New Epoch and the University* (Oration, delivered before the Phi Beta Kappa Society in Sanders Theatre, Cambridge, Mass., June 25, 1896. (2)

14. White, T. W., quoted in Franklin, H. Bruce, ed. *Future Perfect: American Science Fiction of the Nineteenth Century: An Anthology.* Expanded and revised ed. New Brunswick, N.J.: Rutgers University Press, 1995. (132)

15. Engels, Friedrich. *The Condition of the Working Class in Great Britain.* Oxford World's Classics. (36)

16. Mayhew, Henry. *London Labour and the London Poor.* New York: Penguin Classics, 1985. (35)

17. Engels, Friedrich. *The Condition of the Working Class in Great Britain.* Oxford, England: Oxford World Classics. (40)

18. Sharps, Francis, MRCS, quoted in ibid. (162)

19. Ibid. (149)

20. *Illustrated London Times.* Quoted in Heald, Henrietta. *William Armstrong: Magician of the North.* Carmarthen, Wales: McNidder and Grace, 2012. (60)

21. Ibid. (60–61)

22. Leach, Michelle. "5 Disasters that Led to Greater Workplace Safety," *Listosaur: History.* September 2, 2013. (NP)

23. Engels, Friedrich. *The Condition of the Working Class in Great Britain.* Oxford, England: Oxford World's Classics. (245)

24. Ibid. (302)

25. Heald, Henrietta. *William Armstrong: Magician of the North.* Carmarthen, Wales: McNidder and Grace, 2012. (214)

26. Ibid. (215)

27. Ibid. (216)

28. Ibid. (216)

29. Ibid. (217)

30. Quoted in ibid. (219)

31. Quoted in ibid. (220)

32. Ibid. (221)

33. Ibid. (224)

34. Quoted in ibid. (228)

35. Marx, Karl, and Friedrich Engels. *The Communist Manifesto.* The Floating Press, 2008 (from 1888 edition). (5)

36. Ibid. (10)

37. Ibid. (10)

38. Butterworth, Alex. *The World that Never Was: Dreamers, Schemers, Anarchists and Secret Agents*. New York: Vintage Books, 2011. (11)

39. Ibid. (12)

40. Verne, Jules. *Paris in the 20th Century*. Trans. Howard, Richard. New York: Ballantine Books. 1996. (55)

41. Ibid. (55)

42. Ibid. (11)

43. Ibid.

44. Ibid. (30)

45. Ibid. (32)

46. Ibid. (41)

47. Bernstein, Richard. "The New Jules Verne, Like '1984' but Older." *New York Times Book Review*. Dec. 27, 1996.

48. Butterworth, Alex. *The World that Never Was: Dreamers, Schemers, Anarchists and Secret Agents*. New York: Vintage Books, 2011. (189)

49. Ibid. (192)

50. Ibid. (195)

51. Verne, Jules. *Paris in the 20th Century*. Trans. Howard, Richard. New York: Ballantine Books, 1996. (157)

52. Simmons, James R., Jr. *Factory Lives: Four Nineteenth-Century Working Class Autobiographies*. Peterborough, Ont.: Broadview Press, 2007. (24)

53. Ibid. (136)

54. Ibid. (124)

55. Butterworth, Alex. *The World that Never Was: Dreamers, Schemers, Anarchists and Secret Agents*. New York: Vintage Books, 2011. (196)

56. Morris, quoted in ibid. (197)

57. Verne, Jules. *Paris in the 20th Century*. Trans. Howard, Richard. New York: Ballantine Books, 1996. (157)

58. Ibid. (58)

59. Déjacque, Joseph, quoted in *Non-Market Socialism in the Nineteenth and Twentieth Centuries*. Eds. Maximilien Rubel and John Crump. New York: Springer, 1987. (65)

60. Butterworth, Alex. *The World that Never Was: Dreamers, Schemers, Anarchists and Secret Agents*. New York: Vintage Books, 2011. (199)

61. Spies, August, quoted in Butterworth, Alex. *The World that Never Was: Dreamers, Schemers, Anarchists and Secret Agents*. New York: Vintage Books, 2011. (212)

62. Verne, Jules. *Paris in the 20th Century*. Trans. Howard, Richard. New York: Ballantine Books, 1996. (163)

63. Ibid. (164)

64. Ibid. (194)

65. Ibid. (195)

66. Butterworth, Alex. *The World that Never Was: Dreamers, Schemers, Anarchists and Secret Agents.* New York: Vintage Books, 2011. (15)

67. Ibid. (15)

68. Ibid. (27)

69. George, Henry. "The American Republic." *Henry George: Collected Journalistic Writings, Volume 1.* Armonk, N.Y.: M. E. Sharpe, 2003. (113)

70. Butterworth, Alex. *The World that Never Was: Dreamers, Schemers, Anarchists and Secret Agents.* New York: Vintage Books, 2011. (205)

71. Ibid. (209)

**EIGHT: OF ACID AND ACCIDENT**

1. Virilio, Paul. *Pure War* (Interview with Sylvère Lotringer). Trans. Mark Polizzotti. New York: Semiotext(e), 1997.

2. Wright, A. S., and P. J. Maxwell. "Robert Liston, M.D. (October 28, 1794–December 7, 1847): The Fastest Knife in the West End," *American Surgeon,* Jan. 2014; 80 (1): 1–2.

3. Gordon, Richard. *Great Medical Disasters.* Looe, England: House of Stratus, 2001. (3)

4. Ibid.

5. Liston, Robert. "Spontaneous Mortification—Case occurring in the upper extremity." Lecture. *The Lancet London.* Vol. 1. 1835. Google ebook. (848)

6. Lister, Joseph. "On the Antiseptic Principles in the Practice of Surgery," *The British Medical Journal,* Vol. 2, No. 351 (Sept. 21, 1867). (246–248)

7. Ibid. (246)

8. "Medico-Parliamentary," *The British Medical Journal,* Vol. 1, No. 374 (Feb. 29, 1868). (210)

9. Ogston, Alexander. "Case of Carbolic Acid Poisoning," *The British Medical Journal,* Vol. 1, No. 527 (Feb. 4, 1871). (116)

10. "Antiseptic Surgery," *The British Medical Journal,* Vol. 2., No. 990 (Dec. 20, 1879). (1000–1005, 1004)

11. "Antiseptic Ovariotomy," *The British Medical Journal,* Vol. 2, No. 1075 (Aug. 6, 1881). (232)

12. Jenkins, Henry. "Foreword," *Vintage Tomorrows.* Eds. Carrott, James, and Brian David Johnson. Sebastopol, Calif.: O'Reilly, 2013. (xii)

13. Personal interview: Andrew Elis. April 27, 2015. From "Arcanum Inspired Steam-punk Acid Gun," Seller page, TheGOAD. https://www.etsy.com/listing/101905632/arcanum-inspired-steampunk-acid-gun. Accessed April 25, 2015.

14. Schwarcz, Joe. "Vitriolic Attacks," *Office for Science and Society*, McGill Blogs. Accessed April 24, 2015.

15. Karpenko, Vladimír, and John A. Norris. *Vitriol in the History of Chemistry*. Chem. Listy, Vol. 96, 2001. (997–1005, 998)

16. Gaskell, P. *The Manufacturing Population of England: Its Moral, Social, and Physical Conditions*. London: Baldwin and Cradock, 1833. (300)

17. "Execution of Hugh Kennedy." *Loyal Reformer's Gazette 3-4*. Glasgow: Muir, Gowans and Company, 1831. (332)

18. Quoted in Morison, Elting E. *Men, Machines, and Modern Times*. Cambridge, Mass.: MIT Press, 1966. (14)

19. Gaskell, P. *Artisans and machinery: the moral and physical condition of the manufacturing population. considered, with reference to mechanical substitutes for human labour*, London: John W Parker, 1836, A2.

20. Ibid. (v–vi)

21. Tarde, Gabrielle, quoted in Holmes, Joseph. "Crime and the Press," *Journal of Criminal Law and Criminology*, Vol. 20, No. 1 (May 1929). (10)

22. Ibid.

23. Harris, Ruth. "Melodrama, Hysteria, and Feminine Crimes of Passion in the Fin-de-Siècle," *History Workshop*, Vol. 25 (Spring 1988). (38)

24. Ibid. (54)

25. Shapiro, Ann-Louise. *Breaking the Codes: Female Criminality in Fin-de-Siècle Paris*. Stanford, Calif.: Stanford University Press, 1996.

26. Verne, Jules. *20,000 Leagues Under the Sea*. Project Gutenberg. Release date: May 24, 2008 [Ebook #164]. Accessed Aug. 8, 2015. (NP)

27. "Electricity: A Powerful Agent," *Undersea Warfare*, Winter 2004. (NP) Accessed online. July 24, 2015. http://www.navy.mil/navydata/cno/n87/usw/issue_21/verne2.htm.

28. Ibid.

29. Verne, Jules. *20,000 Leagues Under the Sea*. (NP)

30. Morus, Iwan Rhys, "Batteries, Bodies, and Belts: Making Careers in Victorian Medical Electricity." (217)

31. Ibid.

32. Morus, Iwan Rhys. *Shocking Bodies: Life, Death and Electricity in Victorian England*. Stroud, England: The History Press, 2011. (103)

33. Maines, Rachel. *Technology of Orgasm: "Hysteria," the Vibrator, and Women's Sexual Satisfaction*. Baltimore: Johns Hopkins University Press, 2001. (82)

34. Ibid. (85)

35. Morus, Iwan Rhys. *Shocking Bodies: Life, Death and Electricity in Victorian England*. Stroud, England: The History Press, 2011. (107)

36. Lobb, Harry William. *On the Curative Treatment of Paralysis and Neuralgia*. In series *Electrotherapeutics*, London 1859. (13)

37. Ibid. (14)

38. Ibid. (17)

39. Morus, *Shocking Bodies: Life, Death, and Electricity in Victorian England*. Stroud, England: The History Press, 2011. (110)

40. Pye-Smith, Charlie. "The Palace of a Modern Magician," newspaper article cited from: "Cragside Northumberland." *Heritage Group Website for the Chartered Institution of Building Services Engineers*.

41. Ibid.

42. Peterson, J. K. *The Telecommunications Illustrated Dictionary*, 2nd ed. Boca Raton, Fla.: CRC Press, 2002. (732)

43. Fryer, Tim. "Let There Be Light: Early Electricity and the Birth of Modern Luxury Living," *New Electronics*. Sept. 9, 2014. http://www.newelectronics.co.uk/electronics-technology/let-there-be-light-early-electricity-and-the-birth-of-modern-luxury-living/64101/.

44. Larson, Erik. *The Devil in the White City: Murder, Magic, and Madness at the Fair that Changed America*. New York: Vintage Books, 2003.

45. Loyson, Peter. "Chemistry in the Time of the Pharaohs," *J. Chem. Edu.*, Vol. 88, No. 2 (2011). (146)

46. "Justus von Liebig and Friederich Wöhler," Chemical Heritage Foundation. ChemicalHeritage.org. Accessed May 4, 2015. http://www.chemheritage.org/discover/online-resources/chemistry-in-history/themes/molecular-synthesis-structure-and-bonding/.liebig-and-wohler.aspx.

47. Cansler, Clay. "Where's the Beef?" *Chemical Heritage Magazine*. Chemical Heritage Foundation. Fall 2013/Winter 2014. Accessed May 4, 2015.

48. Ibid.

49. Ibid.

50. Finlay, Mark. "Quackery and Cookery: Justus von Liebig's Extract of Meat and the Theory of Nutrition in the Victorian Age," *Bulletin of Medical History*, Vol. 66, No. 3, (Fall 1992). (404–418)

51. Larson, Erik. *The Devil in the White City: Murder, Magic, and Madness at the Fair that Changed America*. New York: Vintage Books, 2003. (35–36)

52. Tremeear, Janice. *Illinois' Haunted Route 66*. Stroud, England: History Press, 2013. (56)

53. Ibid.

54. Ibid. (57)

55. Larson, Erik. *The Devil in the White City: Murder, Magic, and Madness at the Fair that Changed America*. New York: Vintage Books, 2003. (150)

56. Mansfield, Laura, and Elisa Oliver, eds. "A History of Meat." *Feast Journal*. Issue 2. http://feastjournal.tumblr.com/post/103825804347.

57. Ibid. (364)

58. Haigh, J. G. "To Mr. and Mrs. J. R. Haigh." PCOM 8/818 of the National Archives.

59. "John Haigh: The Acid Bath Murderer," *Crime and Investigation*. Web accessed May 8, 2015. http://www.crimeandinvestigation.co.uk/crime-files/john-haigh-the-acid-bath-murderer.

60. Oates, Jonathan. *John George Haigh, The Acid-Bath Murderer: A Portrait of a Serial Killer and His Victims*. Barnsley, England: Wharncliffe, 2014. (viii)

61. Ibid. (193)

62. Somerfield, Stafford. *The Complete and Revealing Story of John George Haigh*. 1950. Quoted in ibid.

63. Oates, *John George Haigh*. (189)

64. Calamai, Peter. "The Real Sherlock Holmes." *Cosmos*. March 14, 2015. Web accessed May 8, 2015.

**NINE: THE SCIENCE OF SHERLOCK**

1. Beattie, J. M. *The First English Detectives: The Bow Street Runners and the Policing of London*. Oxford, England: Oxford University Press, 2012. (1)

2. Ibid. (2)

3. Kenealy, Edward Vaughan. *The Trial at Bar of Sir Roger C.D. Tichborne, Bart: In the Court [. . .] Vol. 7.* Dec. 11, 1873. (157)

4. Blum, Deborah. *The Poisoner's Handbook: Murder and the Birth of Forensic Medicine in Jazz Age New York*. New York: Penguin Publishing Group. Kindle Edition. (1)

5. McDermid, Val. *Forensics. The Anatomy of Crime*. London: Wellcome Collection. Profile Books, Ltd. 2014. (88)

6. Ibid. (90)

7. Morison, Elting. *Men, Machines, and Modern Times*. Cambridge, Mass.: MIT Press, 1966. (125)

8. McDermid, Val. *Forensics. The Anatomy of Crime*. London: Wellcome Collection. Profile Books, Ltd. 2014. (90)

9. Quoted in ibid. (90)

10. Quoted in ibid. (90)

11. Ibid. (91)

12. Ibid. (92)

13. Ibid. (94)

14. Doyle, Arthur Conan. *A Study in Scarlet*. Project Gutenberg. Release date: July 12, 2008 [Ebook #244]. (NP)

15. Ibid.

16. Wagner, E. J. *The Science of Sherlock Holmes*. Hoboken, N.J.: John Wiley and Sons, 2006. (8)

17. Morison, Elting. *Men, Machines, and Modern Times*. Cambridge, Mass.: MIT Press, 1966. (68)

18. Doyle, Arthur Conan. *A Study in Scarlet*. Project Gutenberg. Release date: July 12, 2008 [Ebook #244]. (NP)

19. Quoted in Liebow, Ely. *Dr. Joe Bell: Model for Sherlock Holmes*. Madison, Wisc.: Popular Press. 1982. (2)

20. Doyle, Arthur Conan. *Sign of Four*. Project Gutenberg. Release date: March 2, 2011 [Ebook #2097]. (NP)

21. Burney, Ian, and Neil Pemberton. *Murder and the Making of English CSI*. Baltimore: Johns Hopkins University Press, 2016. (3)

22. Ibid. (5)

23. Ibid. (5)

24. Ibid. (10)

25. Ibid. (11)

26. Burney, Ian, and Neil Pemberton. "Making Space for Criminalistics: Hans Gross and Fin-de-Siècle CSI," *Studies in History and Philosophy of Biological and Biomedical Sciences*, Vol. 44, No. 1 (March 2013). (16–25)

27. Ibid. (16–25)

28. Berg, Stanton O. "Sherlock Holmes: Father of Scientific Crime and Detection," *Journal of Criminal Law and Criminology*, Vol. 61, No. 3 (1971).

29. Quoted in Steenberg, Lindsay. *Forensic Science in Contemporary American Culture*. (35)

30. Ibid. (35)

31. Berg, Stanton O. "Sherlock Holmes: Father of Scientific Crime and Detection." *Journal of Criminal Law and Criminology*, Vol. 61, No. 3 (1971).

32. Locard, Edmond. "The Analysis of Dust Traces, Part I," *The American Journal of Police Science*, Vol. 1, No. 3 (May–June 1930). (276–298)

33. Berg, Stanton O. "Sherlock Holmes: Father of Scientific Crime and Detection," *Journal of Criminal Law and Criminology*, Vol. 61, No. 3 (1971).

34. McDermid, Val. *Forensics*: *The Anatomy of Crime*. London: Wellcome Collection and Profile Books, Ltd., 2014. (67)

35. Ibid. (67)

36. Burney, Ian, and Neil Pemberton. *Murder and the Making of English CSI*. Baltimore: Johns Hopkins University Press, 2016. (91)

37. Burney, Ian, and Neil Pemberton. *Murder and the Making of English CSI*. Baltimore: Johns Hopkins University Press, 2016. (65–66)

38. Savage, quoted in ibid. (66)

39. Ibid. (81)

40. Burney, Ian and Neil Pemberton. *Murder and the Making of English CSI*. Johns Hopkins. 2016, 88.

41. Liebow, Ely. *Dr. Joe Bell: Model for Sherlock Holmes*. Madison, Wisc.: Popular Press, 1982. (2)

42. Jenkins, Henry. *Vintage Tomorrows*, eds. James Carrott and Brian David Johnson. Sebastopol, Calif.: O'Reilly, 2013. (160)

43. Doyle, Arthur Conan. *The History of Spiritualism V. 1*. ReadHowYouWant, Large format ebook, 2008. (2)

44. Falcone, Arcadia. "The Adventure of the Immortal Detective: Discovering Sherlock Holmes in the Archives." *Cultural Compass*. University of Texas Austin. Accessed Dec. 26, 2016.

45. Winter, Alison. *Mesmerism*. Chicago: University of Chicago Press, 1998. (3)

46. Chism, Stephen. "'The Very Happiest Tiding': Sir Arthur Conan Doyle's Correspondence with Arkansas Spiritualists," *Arkansas Historical Quarterly*, Vol. 59, No. 3 (Autumn 2000). (299–310)

47. Seifer, Marc J. *Wizard: The Life and Times of Nikola Tesla: Biography of a Genius*. Secaucus, N.J.: Citadel Press, 1998. (91)

48. Ibid. (92)

49. Tesla, Nikola. *My Inventions and Other Writings*. Courier Dover Publications, 2016. (57)

50. Chism, Stephen. "'The Very Happiest Tiding': Sir Arthur Conan Doyle's Correspondence with Arkansas Spiritualists," *Arkansas Historical Quarterly*, Vol. 59, No. 3 (Autumn 2000). (299–310)

51. Interview, Sir Arthur Conan Doyle. Recorded by William Fox, *Fox Movietone News*, October 1928.

52. Ibid.

53. Ibid.

54. Cervetti, Nancy. *S. Weir Mitchell, 1829–1914: Philadelphia's Literary Physician*. Philadelphia: University of Pennsylvania Press, 2012. (78)

55. Tesla, Nikola. *My Inventions and Other Writings*. Courier Dover Publications, 2016. (58–59)

56. Lamont, Peter. "Spiritualism and a Mid-Victorian Crisis of Evidence," *The Historical Journal*, Vol. 47, No. 4 (2004). (897–920)

57. Interview, Sir Arthur Conan Doyle. Recorded by William Fox, *Fox Movietone News*, October 1928.

58. Ibid.

59. Doyle, Arthur Conan. "Adventure of the Empty House." *Return of Sherlock Holmes*. *The Strand Magazine*, Vol. 26, October 1903. Project Gutenberg. Release date: February 1995 [Ebook #221]. (NP)

60. Mitchell, C. Ainsworth, quoted in *Current Opinion*. Eds. Edward Jewitt Wheeler and Frank Crane. Current Literature Publishing Company, 1911. (280)

61. Mitchell, C. Ainsworth, quoted in ibid. (280)

62. McDermid, Val. *Forensics. The Anatomy of Crime*. London: Wellcome Collection and Profile Books, Ltd., 2014. (290)

### FINISHING THOUGHTS: MAD SCIENCE REPRISE

1. Morison, Elting. *Men, Machines, and Modern Times*. Cambridge, Mass.: MIT Press, 1966. (224)

2. Ibid. (226)

3. Jenkins, Henry. *Vintage Tomorrows*, eds. James Carrott and Brian David Johnson. Sebastopol, Calif.: O'Reilly, 2013. (113)

4. Bowser, Rachel, and Brian Croxall. *Like Clockwork: Steampunk Pasts, Presents, and Futures*. Minneapolis: University of Minnesota Press, 2016. (xxxiv)

5. Ibid. (xxxvii)

6. Gibson, William. Interview with Scott Thill. *Wired*, Sept. 7, 2010. https://www.wired.com/2010/09/william-gibson-interview/.

# INDEX

# INDEX

# INDEX

# INDEX

# INDEX

## N

Nantes, France, 95–96
Napoleon III, 186–188, 192, 195, 233–234, 251
*Narrative of Arthur Gordon Pym, The* (Poe), 104, 105
natural gas, 84, 85
natural philosophy, 25
nerves, 77
Neverwas Haul, xii, 119
New Epoch, 4–5, 59, 92–93, 113, 170, 181–182, 257, 258
*New Epoch, The* (Morison), ix–x, xiii–xiv, xviii
New World, 94, 133
New York, 144
New Zealand, 101
Newcastle, England, 131, 137, 173–174
Newcomen, Thomas, 83
Newton, Isaac, 4, 9–10, 12, 18–30, 32, 45, 62, 63, 116, 119–120
Niagra Power Station, 148, 162–163
Nietsche, Frederick, 30
Niger, 108
*Night of the Bubbling Death*, 218
Nihell, Elizabeth, 52
Nine-Hour Movement, 182–185
"Ninety-Five Theses" (Luther), 9
nitric acid, 214
Noble, Andrew, 183
Nolan, Christopher, 148
Norris, Charles, 225
North Pole, 104, 105–106
northwest passage, 104, 106
novels, 200
nuclear weapons, 155
numbers, 4, 9
numerology, 16–17

## O

Oates, Titus, 105–106
Obeah, 95
O'Hare, Shannon, xii
O'Neill, James, 151
operating rooms, 204–205
*Opticks* (Newton), 4
order, 18, 21, 29, 55
Orfila, Mathieu Joseph Bonaventure, 233–236
organic chemistry, 219–220

## P

Padua, Sydney, 126, 128, 140
Palmer, Arthur, 143
paradigm shift, 8
Paris, 187, 195–196, 211
Paris Exposition of 1864, 187, 195
*Paris in the 20th Century* (Verne), xvi–xvii, 187–188, 191–195
Paris World's Fair, 196
Park, Mungo, 92, 108–111
past, xii, xiii, xv, 80, 259
Pasteur, Louis, 202–203, 204
patents, 84, 144
pathologists, 232
pathology, 243–244
*Pattern Recognition* (Gibson), 258
Pemberton Mill, 180–181
pests, 5
Peters, S. M., 10–12, 29, 41–42, 258
petroleum, 52
phenol, 201–207
Philippe (king), 186, 187
philosophy
    mechanistic, 33–34
    natural, 25
Phipps, George Augustus Constantine, 215
photography, 254
physicians, 231–232
physio-mathematics, 46
pig iron, 234
Pilâtre de Rozier, Jean-François, 76
Plana, Giovanni, 124
planetary movement, 14–16
planets, 7, 14–15, 80
Planté, Raymond-Louis-Gaston, 217
plastics, 52
Poe, Edgar Allen, 104
poisons, 95, 232–235
police, 231. *See also* detectives
pollution, 136, 249–250
Pope, Alexander, 29
postmodern era, xvi
poverty, 178
power, x, xiii, 11, 61, 82, 84, 92–93
    electric, 61, 134, 141, 166–167
    engineering, 118
    hydraulic, 61, 134, 137, 141–142, 217
    manufacture of, 116, 118, 134

296

# INDEX

# ACKNOWLEDGMENTS

F irst: To the Dittrick Medical History Center and Museum, James Edmonson, Jennifer Nieves, and Laura Travis, thank you. History owes a deep debt to all museums, everywhere.

To my colleagues, family, and friends, who lent advice and support, my deep gratitude. To readers of early drafts, and most especially Lance Parkin, I owe you—and you know where I live. To Roger Whitson and to all those who kindly submitted to interview questions, endless thanks. And finally, to Mark, who patiently listened to more drafts with more encouragement than any other human, all my love. Also, you are stuck with me.